THE SECULAR ARK

JANET BROWNE

The Secular Ark

Studies in the History of Biogeography

Yale University Press
NEW HAVEN AND LONDON

Published with assistance from the foundation established in memory of Amasa Stone Mather of the Class of 1907, Yale College.

Illustrations 1, 2, and 4 are reproduced by courtesy of the Wellcome Trustees, illustration 3 and figure 1 by courtesy of the British Museum (Natural History). The table in chapter 8 is a version of the author's own table and is reproduced with the permission of the *Journal of the History of Biology*. All other figures are printed from photographs kindly supplied by the Science Museum Library, London.

Designed by James J. Johnson and set in Caledonia Roman type by Graphic Composition, Inc. Printed in the United States of America by Murray Printing Company, Westford, Mass.

Library of Congress Cataloging in Publication Data

Browne, Janet.
The secular ark.

Bibliography: p.
Includes index.
1. Biogeography—History. I. Title.
QH84.B78 1983 574.9 82-17497
ISBN 0-300-02460-6

10 9 8 7 6 5 4 3 2 1

Contents

Illustrations

ILLUSTRATIONS

FIGURES

Preface

I had not intended to introduce this volume, believing with Jonathan Swift that one can read too many prefaces. But I would not like to mislead readers into thinking that this was a complete history of biogeography—rather, it is a series of incidents and themes that interested me and seemed to make an unusual story. In particular, I must point out that I have not considered in any detail continental European ideas after 1855, when Alphonse de Candolle published his magisterial *Géographie botanique raisonnée*. There was, of course, an immense interest in plant and animal distribution patterns after that time from scholars outside Britain, and major theoretical analyses were being put forward elsewhere, especially in Germany.

My other point is somewhat more technical in that it has been difficult to avoid using certain words that have special meanings in contemporary science: the collective nouns for plants, such as "association," for example, are used here without any of their modern phytogeographical connotations. The same might be said about the expression "geographical distribution," by which I mean a constellation of ideas about plant and animal topography, not the process of organic dispersal.

Many people have helped me in the course of writing this book and it is a pleasure to be able to acknowledge them in public. Marie and Rupert Hall, in particular, have followed its progress almost as long as I have and have offered their friendly encouragement on countless occasions. During trips to Cambridge I also enjoyed the advice and hospitality of Sydney Smith and Peter Gautrey, and have further benefitted from discussions with David Kohn. To these generous Darwin scholars I extend my warmest thanks. A special debt is owed to Professor Evelyn Hutchinson and Stan Rachootin who introduced me to Yale University Press. Their separate comments on my thesis were greatly appreciated. No less a debt is due to Bill Bynum of the Wellcome Institute for the History of Medicine, who has affably given me all the time I needed to finish this book and has quietly taken on many tasks that I ought to have done. It has been a delight to work with him and the rest of the Wellcome staff.

Other friends and scholars have helped me by commenting on parts of the manuscript at one stage or another: I am particularly grateful to Dov Ospovat, Sam Schweber, Joy Harvey, Martin Rudwick, David Roos, David Kohn, and Debbie Dwork. Malcolm Nicolson valiantly read it twice and I thank him for his valuable suggestions. More generally, I have had the advantage of discussions with colleagues at the Wellcome Institute, with the one-time Imperial College history of science seminar group, and with members of the Society for the Bibliography of Natural History. Parts of chapters 1 and 3 were presented to meetings of the British Society for the History of Science (1980) and the Joint Atlantic Seminar in the history of biology (1979), respectively, and most of chapter 8, on Darwin's principle of divergence, was first published in 1981 in the *Journal of the History of Biology*. Comments on these papers were exceptionally useful. Any historical gaffes remain, of course, my own. The Syndics of Cambridge University Library, the Archives of Imperial College, the Library of the British Museum (Natural History), and the Library of the Royal Botanic Gardens, Kew, have courteously given permission to quote from manuscripts in their possession; I particularly wish to thank the staffs of these institutions for their unfailingly helpful and friendly assistance. And although they both had things they would far rather have done, Nick Browne and Heather Edwards generously gave up most of their summer to type the final manuscript.

Most of the research for this book was carried out with the gratefully acknowledged aid of a Major State Studentship from the Department of Education & Science and the equally appreciated support of a Keddy Fletcher Warr Scholarship from the University of London. My work on Joseph Hooker was also helped by a timely Royal Society grant for research in the History of Science. One way or another, the Wellcome Trustees have supported me since 1979, and I am pleased to acknowledge their generosity.

ONE

Descent from Ararat

Six months before setting off on the expedition that was to culminate in his death at Khartoum, General Gordon was on leave in Palestine, Bible in hand, immersed in a prolonged series of researches into biblical history and topography. He had never been happy with the idea of Mount Ararat as the Ark's final resting place and was now methodically investigating the claims of other sites being the scene for this miraculous event. A firm believer in the literal truth of the Bible, with slight geographical revisions, Gordon was, however, sadly out of step with his contemporaries. Most eminent Victorians had already adopted a higher form of biblical criticism, with the comparison of religions and a hard-headed, even cynical evaluation of theological symbols. The Bishop Colenso affair was over; George Eliot was dead; *Jude the Obscure* was soon to come. Doubt rather than unquestioning faith was the prevailing attitude to the church and its written records, to the Bible and the book of Genesis.

For most scientists of the 1880s there was no question of a literal interpretation of the Noachian Deluge. Charles Darwin, for one, had treated the idea of an Ark with derision, telling his friend the geologist, Charles Lyell, that the giant mastodon must have perished in the waters because it could not squeeze through the door. Alfred Russel Wallace, Joseph Hooker, Louis Agassiz, and Edward Forbes had all agreed on the implausibility of Genesis as a true account of the world's early history. Indeed, they assumed that the legend of Noah and his Ark had materially impeded the development of more acceptable ideas about the animal kingdom and believed that it was not until the discrediting of the Ark toward the middle of the eighteenth century that modern zoology began to flourish.

But these self-consciously "modern" naturalists were forgetting—or, more likely, were completely unacquainted with—the critical role played by Noah's Ark in the development of ideas about the geographical distribution of animals and plants. The idea of an Ark in which pairs of animals were preserved during the Deluge had been a concept of far-reaching significance, as had the disembarkation on Mount Ararat and the subsequent dispersal of animals over the unoccupied globe. The biblical story, in fact, had done a great deal to stimulate investigations into the natural world and, among other things, provided the first systematic explanation for the phenomena of biogeography. Far from being the intellectual impediment ridiculed by Darwin and his circle, the theological certainty that Gordon anxiously searched for, the idea of an Ark focussed scholarly attention on the topographic arrangements of species, as well as encouraging naturalists to build up a repertoire of theoretical commitments and practical expertise in the analysis of organic distribution. The Ark actually promoted, rather than hindered, the growth of this particular branch of the natural historical sciences.

It would be misleading to suggest, however, that the notion of Noah's Ark itself was the starting point for geographical studies, for few things in history really "start" in such an emphatic manner. Even the most cursory glance at the history of natural history reveals a wealth of individuals interested in the general problem of animal and plant distribution. From the earliest times mankind seems to have been aware of differences in living beings according to country and to have placed a value on animals and plants not found in his own region. As empires were founded and trade routes established, so knowledge of the natural world grew, goods and indigenous products were bought and sold, animals and plants were exchanged and introduced to new environments, and a substantial body of practical information about the variety of living things and their geography became progressively available. Clearly, an awareness of distribution patterns came into being independently of the Bible.

Indeed, it was only with the Renaissance and the subsequent change in attitudes toward Genesis that Noah's Ark took on an influential new role in European philosophy. Gradually, over the sixteenth century, the Scriptures lost much of their metaphorical and allegorical meaning and came to be understood by many as a literal tract, a real account of the origins of the natural world. The Bible, it was argued, was grounded in fact rather than imagery. With such an emphasis on literalism, the story of Noah—indeed all the patristic writings—acquired a new importance; and to explain and justify the Mosaic account of creation and the Deluge became an act of the highest theological significance. In one sense, a literal interpretation of the Bible would help theologians reaffirm the power and glory of God; in another sense, a successful account would dramatically confirm the Christian church's authoritative position in contemporary society. Late-sixteenth-century churchmen (who were not unaware of these possibilities, of course) therefore encouraged one another to take up the new hermeneutics—to consider biblical texts as if they described real events.

The churchmen's advice was generally heeded—although Sir Thomas Browne doubted that the Scriptures were solely literal writings—and by the middle of the seventeenth century the practice of expounding the Bible's reality was widely established. Protestants[1] and Catholics alike directed their attention to the story of Noah. Soon, the literal Ark was every theologian's favourite field of inquiry.

THE LITERAL ARK

Among the theological scholars most intrigued by the Ark was the German Jesuit priest Athanasius Kircher (1602–1680). Like many Jesuits, Kircher displayed a marked taste for natural philosophy, particularly the exotic and arcane, and devoted his entire intellectual life to the publication of more than thirty books crammed with esoteric learning and experimentation, ranging from the study of languages and hieroglyphics to music, magnets, and microscopes.[2] Curious and encyclopedic in his learning, enthusiastic and flamboyant in his lifestyle, he ended up as a professor of mathematics at the Roman College of the Society of Jesus, the greatest centre of Jesuit scholarship, in Rome in 1634. There he stayed until his death in 1680, teaching and writing, maintaining a vast, unwieldy correspondence with natural philosophers throughout Europe, including Henry Oldenburg, the Secretary of the Royal Society.

The focus of Kircher's life, however, was not simply zest for the natural sciences, but also a strong, uncomplicated religious faith that provided the motivation for his scientific activity. He was staunchly committed to the idea that science would reinforce Christian doctrines. Natural philosophy did not seduce the faithful from the Church, it actually strengthened established religion by demonstrating how

God's work could be explained in terms that mere human beings could understand. To Kircher, the Bible was more than an inspired tract, the symbol of another, supernatural world revealed through imagery alone. He saw it rather as a record of real events. The Old Testament and the Mosaic account of creation were literally true, the story of Noah and the Deluge a potent description of the first great salvation after the fall of man. In 1675 Kircher thus took it upon himself to prove that Noah's Ark—second in theological importance only to Adam and Eve—was entirely comprehensible in the light of contemporary natural philosophy. Making full use of recent developments in historical, literary, philological, and scientific studies, he thereby undertook one of the first tentative surveys of the primitive history of the earth and its organisms.

Like other Scriptural scholars, Kircher first sought to answer the old question about the size and shape of the Ark. Was it big enough to accommodate all the animals? Kircher minutely scrutinised the dimensions given in Genesis. Unlike other biblical commentators, he concluded that Noah must have used the common cubit (one and one-half feet) for his calculations, not the larger geometrical unit. If Noah allowed for adequate aisles and passageways, using the common cubit, then three hundred animal stalls could be built on the lower deck, each big enough to house an elephant. The second floor provided a great storage hold, containing food enough for a vast city, and the top deck, according to Kircher, must have held at least two hundred bird cages and the accommodation for the human passengers (illustration 1).

Since Kircher listed no more than 130 species of animals, 30 pairs of snakes, and 150 different kinds of birds, there was obviously no overcrowding. Nor did Noah need to take fish on the Ark, or those reptiles and insects that spawned from putrefaction. Indeed it was not even necessary to take hybrid animals, in Kircher's view, or other species that had been subsequently altered by climate or had degenerated from their original created state. With all these exclusions the number of species was small enough for a moderate-sized Ark.[3]

But criticism of Kircher's account soon mounted. His book was merely one in a continuing series of biblical commentaries, and scholars had already begun to question the central issues he had covered so authoritatively. The size of the Ark could not be dismissed out of hand. Many people reasserted that Moses could only have used geometrical cubits to describe the Ark's construction, rather than common cubits one-sixth their size. And since these measurements may have referred to only one deck out of three, they ought to be tripled for the final dimensions of the vessel. Noah's daily life was endlessly examined and the logistics of stabling, feeding, and cleaning the animals were worked out in exhaustive detail. Other questions of a literal nature arose over provisions for the livestock. What did they eat and where was the fodder stored; did Noah bring extra sheep and goats for the carnivores; what of the disposal of dung? Two extra decks were added to the three described in the Bible to accommodate additional food supplies, to make room for gangways and workrooms, and to house all the domesticated beasts not mentioned by Moses. Additional partitions and stables were introduced to keep unclean animals away from the rest and to provide ventilated living quarters for Noah and his family.[4] Not surprisingly, the proliferation of such ad hoc alterations ultimately transformed the traditional silhouette of the Ark. Where earlier writers had thought of it as pyramidal in shape, Kircher and others believed it to be more like a rectangular house perched on an invisible keel: Kircher's *Arca Noë* of 1675, for example, has something of the appearance of a shoebox or modern hotel. The iconography of this transformation is striking and appears to bear no relation to improvements in contemporary ship design[5] (illustration 2).

Kircher and his generation seemingly had good reason to worry about the actual size of the Ark. For just at this point—just when belief in the literal truth of the Bible was at its highest—foreign travellers began to bring back to Europe an unprecedented wealth of unusual creatures from other parts of the world. To the enormous mid-century catalogues of Ulysses Aldrovandus, Johannes Jonston, and Edward Topsell were added many, more comprehensive lists of regional faunas. Strange species like the kangaroo, anteater, sloth, and dodo were sighted in distant countries and reported to European naturalists.[6] Indeed, by the end of the seventeenth century Britain's premier systematist and natural philosopher, John Ray, knew of over 500 species of birds, 150 quadrupeds, and approximately 10,000 invertebrates of one kind or another. Plainly, there was an abundance of previously unknown forms of life and, unless alternative suggestions could be made, all these had to be enrolled on the Ark's list of passengers. Equally plainly, the Ark was not infinite in size. Literal interpretations of the Bible therefore introduced a dilemma of a magnitude unheard of in previous centuries: all animals had to get into the Ark, yet there were far too many species for this to be remotely possible.

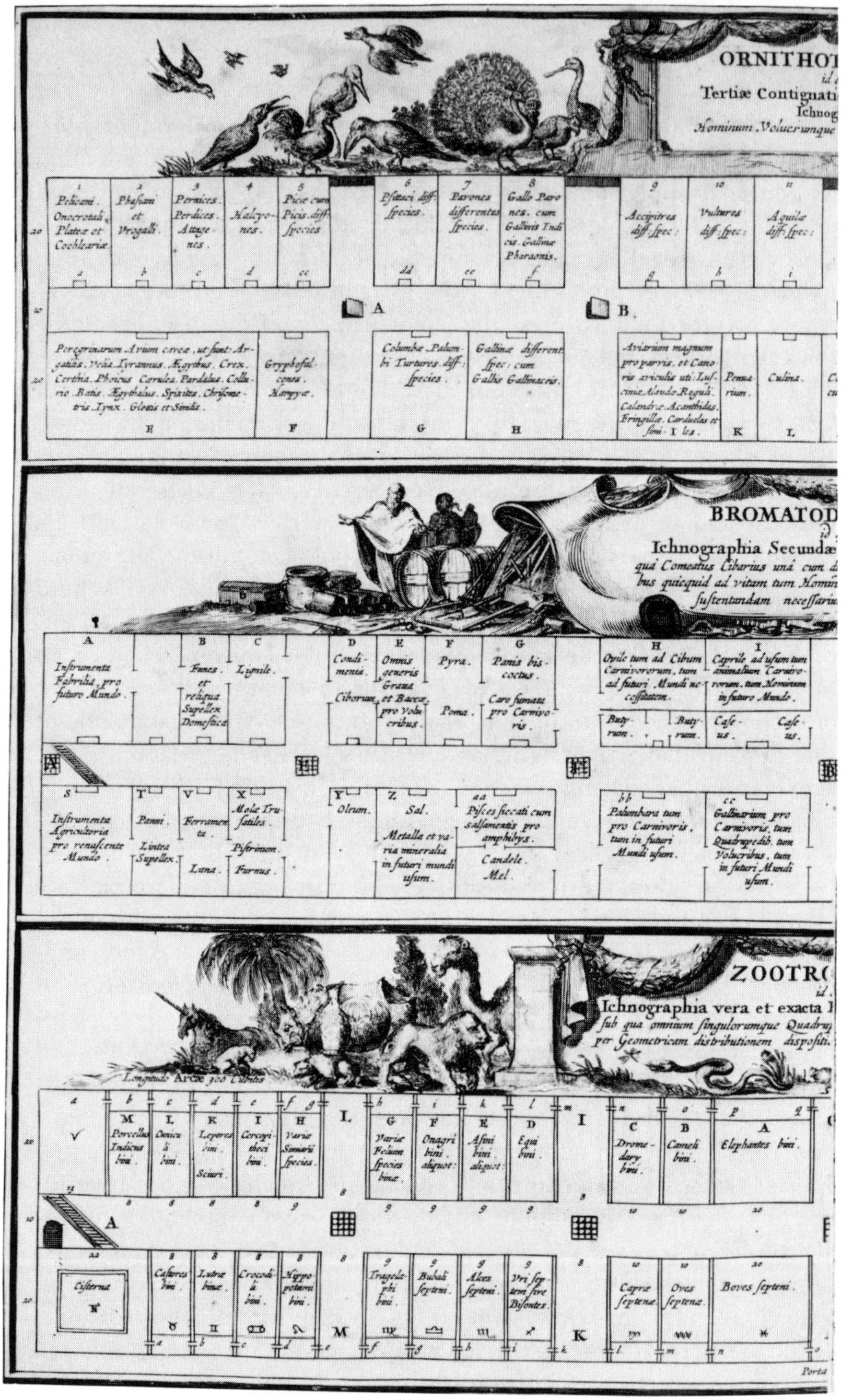

ORNITHOT
Tertiæ Contignati
Ichnog
Hominum Volucrumque
Pelicani. Onocrotali Plateæ et Cochleariæ.
Phasiani et Vrogalli.
Pernices. Perdices. Attagenes.
Halcyones.
Picæ cum Picis diff. species
Psittaci diff. species.
Pavones differentes species.
Gallo Pavones cum Gallinis Indicis. Gallinæ Pharaonis.
Accipitres diff. spec:
Vultures diff. spec:
Aquilæ diff. spec:
A
B
Peregrinarum Avium caveæ, ut sunt: Argatilis. Velia. Tyrannus. Ægythus. Crex. Certhia. Phoicus Cærulea. Pardalus. Collurio. Batis. Ægythalus. Spixites. Chrisometris. Iynx. Glottis et Similia.
E
Gryphosfalcones. Harpyiæ.
F
Columbæ Palumbi Turtures diff: species.
G
Gallinæ different: spec: cum Gallis Gallinaceis.
H
Aviarium magnum pro parvis, et Canoris aviculis uti: Luscinia, Alaudæ Reguli. Calandræ. Acanthides. Fringillæ. Carduelas et simi-I les.
Penuarium.
K
Culina.
L
BROMATOD
Ichnographia Secundæ
quâ Comeatus Cibarius unâ cum
bus quicquid ad vitam tum Homi
sustentandam necessariu
A
Instrumenta Fabrilia, pro futuro Mundo.
B
Funes. et reliqua Supellex Domestica
C
Lignile.
D
Condimenta.
Ciborum
E
Omnis generis Grana et Baccæ pro Volucribus.
F
Pyra.
Poma.
G
Panis biscoctus.
Caro fumata pro Carnivoris.
H
Ovile tum ad Cibum Carnivororum, tum ad futuri Mundi necessitatem.
I
Caprile ad usum tum animalium Carnivororum, tum Hominum in futuro Mundo.
Butyrum.
Butyrum.
Caseus.
Caseus.
S
Instrumenta Agricultoria pro renascente Mundo.
T
Panni.
Lintea Supellex.
V
Ferramenta.
Lana.
X
Molæ Trusatiles.
Pistrinum.
Furnus.
Y
Oleum.
Z
Sal.
Metalla et varia mineralia in futuri mundi usum.
aa
Pisces siccati cum salsamentis pro amphibys.
Candelæ.
Mel.
bb
Palumbaria tum pro Carnivoris, tum in futuri Mundi usum.
cc
Gallinarium pro Carnivoris, tum Quadrupedib. tum Volucribus, tum in futuri Mundi usum.
ZOOTRO
Ichnographia vera et exacta
sub qua omnium singulorumque Quadru
per Geometricam distributionem dispositi
Longitudo Arcæ 300 Cubitis
M
Porcellus Indicus bini.
L
Cuniculi bini.
K
Lepores bini. Sciuri.
I
Cercopitheci bini.
H
Variæ Simiarii species.
L
G
Variæ Felium species binæ.
F
Onagri bini aliquot:
E
Asini bini aliquot:
D
Equi bini.
I
C
Dromedary bini.
B
Cameli bini.
A
Elephantes bini.
A
Cisterna
N
Castores bini.
Lutræ binæ.
Crocodili bini.
Hippopotami bini.
M
Tragelaphi bini.
Bubali septeni.
Alces septeni.
Vri septem sive Bisontes.
K
Capræ septenæ.
Oves septenæ.
Boves septeni.
Porta

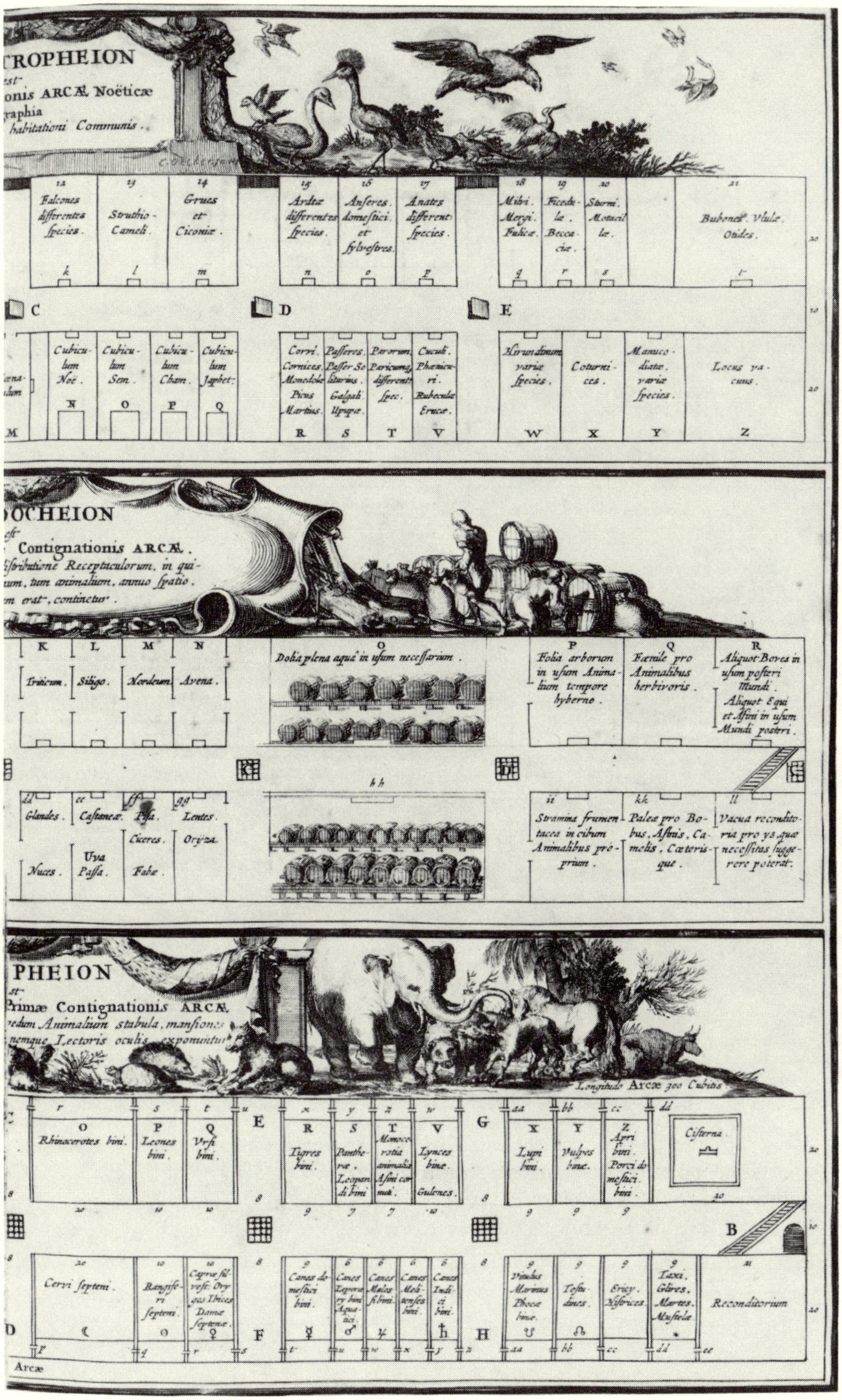

Illustration 1 Inside Noah's Ark. Athanasius Kircher's interpretation of the internal arrangements of the Ark. Literal reconstructions such as this even included gangways, stairs, and ventilation hatches. Animal stalls were arranged in a sequence thought to make feeding and cleaning easier for Noah and his family. From A. Kircher, *Arca Noë* (Amsterdam: Waesberg, 1675), p. 108.

Illustration 2 Loading the Ark. Kircher's external view showed the Ark as a large rectangular box, not the traditional pyramid. Pairs of animals gathering around the entrance reflected his belief in the story of Mount Ararat. From A. Kircher, *Arca Noë* (Amsterdam: Waesberg, 1675), p. 122.

But although many seventeenth-century naturalists, particularly those in Britain like Ray or Tancred Robinson, recognised that the Ark was overloaded, they did not confront the dilemma directly. In fact, they avoided it completely. It is hard to say, in retrospect, whether this was a deliberate and considered policy, designed to circumvent the possibility of being forced into making alternative, nonbiblical suggestions, or whether, like so many other issues in the history of science, the overloaded Ark was obscured by more obvious, more pressing, or more answerable questions in natural philosophy. In the event, however, naturalists for the most part bypassed the Ark and turned instead to the Flood. By arguing that the entire earth had been submerged to a depth of thirteen cubits or more, as specified in the Bible, it was possible to assert that all modern species *must* have been passengers or they would not be in existence today. Living beings not represented in the Ark would inevitably have perished. Since belief in a worldwide flood required all present-day species to have been inside the boat, it was only man's limited imagination that prevented him from seeing how it could be done. So, by turning the argument on its head and reaffirming that the Deluge had covered the whole world, naturalists successfully managed to avoid what appeared to be an exceedingly awkward question about the physical size of the Ark.

Apart from this clever semantic evasion, there was another, more practical reason for a change of emphasis onto the Flood in preference to the Ark itself. There seemed to be tangible evidence that a great mass of water once had existed across the surface of the earth, and this evidence—unlike the proliferation of animals—proved no embarrassment to the biblical story. Fossil sea shells could be found on mountain slopes and often appeared in beds far removed from the existing sea. Other animal remains suggested organisms that had perished under water. To be sure, there were doubts about the supposed organic nature of these "formed stones." In Britain, Martin Lister and Edward Lhuyd took the view (though not without misgivings) that fossils were sports of nature produced by some "plastic virtue" in the earth. Robert Hooke, on the other hand, accepted that they were the remains of real plants and animals and sought an explanation for their presence in the earth's crust. He thought they must be the remnants of a prediluvial world that had been wiped out by a catastrophic and universal deluge. The Flood itself could have been brought on by a rearrangement of the earth's surface through a series of devastating earthquakes.[7]

In 1669 Hooke's arguments received distinguished support from the publication in Italy of Nicholas Steno's *De Solido intra Solidum naturaliter Contento Dissertationes* (A dissertation concerning a solid contained in a solid). Steno undertook to reconstruct the successive transformations that the surface of the earth had undergone in Tuscany. It was clear to him that most of the beds he was dealing with had been deposited by sedimentation—that they had been laid down under water. By supposing that the primitive earth had once been entirely covered with water, deep enough to cover the elevated plateau he knew best, Steno declared that organic remains must have been deposited when the original terrain was submerged. Such an account agreed with the Genesis description. The formation of the world in the bosom of the waters, the emergence of dry land, and its subsequent flooding by a universal sea were all incontrovertibly attested by the Scriptures.

However, this emphasis on the universality of the Flood forced scholars to consider another aspect of the Noachian story that had previously been ignored—namely, the way the earth was restocked with animals and plants. If the Deluge was to be taken literally, all species had originally disembarked from the Ark on Mount Ararat and had dispersed from there across the deserted globe. Literal-minded natural philosophers were therefore obliged to explain how animals had arrived at their final destination. And this, in a sense, marks the first beginnings of the study of geographic dispersal.

DISPERSAL FROM A SINGLE CENTRE

Taking their cue from what was already known about species introduced to Europe, most seventeenth-century commentators supposed that plants and animals had a marked capacity for migration; that they could establish themselves successfully in a new country and, in the case of some particularly exotic species, could even hybridise or degenerate into related forms. Plain observation told men like James Parkinson and the two John Tradescants, father and son, that many imported plants were capable of growing in English gardens. Parkinson's delightful book *Paradisi in Sole* (1629), an elaborate pun on his name Park-in-the-sun, was a hymn to the wide range of naturalised and exotic species available to contemporary gardeners, while the Tradescants made their name by introducing plants such as the Virginia creeper and Chinese lilac into the British Isles. It was clear that the majority of plants could live in countries other than their own, and practical experience of the animal kingdom confirmed this. The black rat had come to Britain on board ship in the Middle Ages; the rabbit was first imported into the same country for its fur and meat in the twelfth century; both species had become notorious pests by the end of the Tudor period. In Ireland, the Fellows of Trinity College Dublin deliberately supplemented the native fauna by introducing common frogs into the ditches of the University Park in 1696.[8] Extremely successful in its new environment, *Rana temporaria* remains after 300 years the only species of frog in Ireland—living proof, as it were, of the capacity of animate beings to flourish in a different region. Such knowledge can only have reinforced the idea that in the Beginning—at the start of the world—animals and plants had migrated widely. In this light, the notion of dispersal from a single centre presented few intellectual problems.

There was, moreover, an even more potent model in the story of Noah's Ark itself. The tenth and eleventh chapters of the book of Genesis told how all men were descended from Noah and how they had made their way from Armenia to their present countries. The story was one of multiplication and diffusion under the guiding hand of God. Although philologists, antiquaries, and philosophers wrangled over the origins of various ethnic groups (tracing them back to one or another of Noah's grandsons), the general picture of a body of people moving out from a spot in Armenia to populate all of the known world was easy to envisage: the land routes were open and the water routes were short. For this reason, seventeenth-century authors believed that direct connections could be traced from each race of human beings back to Noah's three sons, Shem, Ham, and Japheth. The pioneer philologist Edward Brerewood, for example, claimed that the African Negro must have been the descendent of Ham—the one son who fell out of favour and was destined to die cursed. If a direct lineage could not be established Brerewood supposed that one race was descended from another. He used this idea to explain the origins of native American Indians from an ancient Tartar stock that had migrated overland from Armenia.[9] One of Brerewood's more famous contemporaries, Hugo Grotius, took issue with the Tartar theory and proposed that Americans had come instead from Norway, by way of Greenland.[10] Grotius was attacked in turn by the Flemish geographer Laetius, who believed that they had come from Persia.[11] Other scholars, such as Hornius, favoured successive invasions of Phoenicians, Chinese, and

Scythians (all directly related to Noah's sons), each occupying the New World in turn, only to degenerate finally and interbreed with each other.[12] Though the details varied, the underlying assumption was the same for all these theories: there had been a single physical and temporal origin for man, and his present distribution over the globe had been brought about by migration from that one spot. Evidently these views served to reinforce and vindicate the idea that all animals had a common origin from which they too had dispersed.

In explaining the course of animal dispersal, scholars displayed both the weaknesses and strengths of seventeenth-century natural history. Though based on ancient legend, their well-meaning attempts to justify Noah's Ark showed consummate skill in assessing the means of migration, population statistics, and reproductive rates and demonstrated a fine appreciation of the distinctive features of the natural world. José de Acosta, for instance, had listed many species that were to be found only in the Americas.[13] Could they have been carried there by boat with the first human inhabitants of that country? If one accepted that men and animals were created at the same time, and that human and earthly history were broadly synchronous, the suggestion is perfectly reasonable. But some literalists were quick to disagree. Lipsius, for one, took the view that men would hardly travel willingly with rattlesnakes and bears. He went on to conclude that Africa was once connected to the New World by the continent of Atlantis, and that animals had travelled west on dry land all the way.[14]

Most of the experts of Lipsius's generation contented themselves with less extravagant hypotheses about the arrival of animals in America. Athanasius Kircher was reluctant to admit that the surface of the earth had undergone such dramatic changes after the Flood and thought it more likely that animals swam from island to island, or were brought by boat. Some species, he suggested, may have escaped from the ships that were transporting them to princely menageries. Hornius, who favoured the overland migration theory, believed that they crossed in winter, when the northern seas were frozen.[15] Sir Matthew Hale, on the other hand, attempted to explain the American fauna through a kind of migration and subsequent degeneration, the latter being stimulated by the effects of hybridisation, mutation, cultivation, or perhaps the influence of local environmental factors. If all these fail as explanations, he added, there is always the possibility that American species are spread worldwide and that when Africa and Asia are more thoroughly explored it will be found that the New and Old Worlds carry roughly the same populations.[16]

At least one natural philosopher was not satisfied with this manner of special pleading. The French intellectual and diplomat Isaac de La Peyrère (1594–1676) doubted that organisms could migrate as widely as the current philosophy suggested. He realised that such arguments were meaningless without the unwritten assumption that the Flood was universal. If the earth had indeed been covered with water then it was possible—though in La Peyrère's opinion difficult—to believe that Mount Ararat was the source of all postdiluvial species. But what if the Deluge had been a local event, confined to Europe and the Middle East? The Noah story would then refer only to the Biblelands. Elsewhere the various races of man could have lived on, accompanied by a full complement of birds, beasts, and plants, untroubled by the catastrophe.

La Peyrère championed this view in 1655.[17] He was particularly concerned to demonstrate that Adam was not the first-created man, but merely the first of the Jewish race. Adam, he thought, had been preceded by many nations, living in China, America, Greenland, and the mysterious continent to the south. These nations were neither destroyed by the Flood nor involved with the rest of the biblical story. The history of Genesis in fact could be interpreted much more liberally and was in this way "wonderfully reconciled" to the evidence of the natural world.

Such arguments, of course, were exceptional in Catholic France and La Peyrère's *Praeadamitae* (Men before Adam) was roundly condemned by the Paris parliament of 1655. Ostracised by the capital, La Peyrère then fled to Brussels, where he was arrested in 1656 on the order of the archbishop of Malines. On his release he was escorted to Rome to sign a public retraction in the presence of Pope Alexander VII, later ending his days in a seminary near Paris. His early iconoclasm had served only to make him a notorious figure, the source of theological uproar. Yet La Peyrère's attitude to Genesis is clearly important in the context of seventeenth-century thought. In his view there was no need to fit large numbers of animals into the Ark. There was no need to suppose that animals migrated from Ararat or that every human being was derived from Noah and his sons. Using the same means of literal inquiry as his contemporaries, La Peyrère managed to support a thesis fundamentally opposed to conventional beliefs. And although his arguments were exclusively applied to the origins of humankind, it seems that the way was now marked for more secular interpretations of the early history of other living beings. One perceptive naturalist, the Dutchman Abraham van der Myl, recognised this signpost for what it was and seized the opportunity to assert that the animals and men of the New World had lived there from the start and had not arrived by migration from the Middle East. For van der Myl, there had been not one but many creations.[18]

Deeply disturbed by La Peyrère's writings, Sir Matthew Hale suggested in 1677 that the Frenchman's theory was like Pandora's box, revealing a series of theological problems more perplexing even than those faced by the Ark literalists. What, for example, would be the theological status of pre-Adamite man; had he too been divinely created; would he also be saved? La Peyrère's argument, moreover, stipulated that the earth had had no beginning and was in no danger of coming to a dramatic end. Without the twin ideas of a creation and an eventual day of judgement, the Christian faith lost two of its most vital doctrines: La Peyrère's view made nonsense of the central tenets of both the Catholic and the reformed Protestant Churches.

So, in a book on *The Primitive Origination of Mankind*, Hale attempted to refute these "immoral and irreligious principles" by proving that man must have originated with Adam.[19] There, he introduced a line of reasoning that not only endorsed the rejection of La Peyrère's conjectures about Adam, but also remained central to theoretical natural history for sixty years, until the time of Linnaeus, when it was codified in the species concept. Hale's point was simple and effective. Through the normal processes of reproduction, human beings increase in number with every generation. A family of two, he calculated, would grow to eight over a period of thirty-four years.[20] Looking back, it was obvious that every generation was smaller than the one which followed. Logically, this process of diminution must end in a single couple—Adam and Eve—for whom there could be no natural origin. Here, said Hale, one must assume a Creator. The same argument applied to the animal kingdom: all natural groups of animals tended to increase in number, indicating that the species had begun with only two representatives. After God, only Noah could have provided the modern world with two of everything. The Mosaic account, therefore, could be thoroughly vindicated.

Hale's argument certainly eclipsed La Peyrère's theory of pre-Adamite man and the eternity of the world. But there was still room for doubt about the second half of his unconventional thesis, the physical extent of the Flood. Bishop Stillingfleet agreed with La Peyrère that the Flood may well have been a local event.[21] This controversial view was given wider circulation by Matthew Poole in a compendium of biblical criticism published in 1669.[22] The same suggestion reiterated by the French Protestant theologian Jean Le Clerc in his influential series of biblical commentaries[23] inspired a frenzy of diluvialism in Britain. Historians of geology know only too well that British natural philosophers at the end of the seventeenth century were intent on arguing against these provocatively liberal interpretations of Genesis and concentrated on proving the case for a universal Deluge.[24] Thomas Burnet, for example, explained in 1680 how the main events of earth history were the Creation, the Deluge, and the final conflagration, illustrating these stages (including a splendid watery sphere) on the frontispiece of his *Telluris Theoria Sacra* (Sacred theory of the Earth). The Deluge was a necessary event in this scheme because it destroyed the pristine, paradisiacal world that had proceeded directly from the Almighty's hand. By the simple operation of the laws of nature the first world was destroyed and the postdiluvial earth formed out of the ruins. But John Woodward objected to Burnet's belief that all subsequent change on the surface of the earth had been a decline or decay from original perfection, and in 1695 he forcibly denied that these alterations were haphazard and unmeaning. The Deluge, he thought, was not simply a punishment for man's wickedness, but a complete refitting of the globe to adapt it to the needs of fallen creatures. God had provided a world suitable for His creations, whatever their state of grace.[25]

Woodward's views on the religious significance of a worldwide flood had the support of John Ray and William Whiston: they too believed that the Deluge had been a real event, although they differed radically on the source of the diluvial water. Ray, like Woodward and Burnet, thought that the skin of the earth had in some way ruptured to allow vast reservoirs of subterranean water to rise to the surface,[26] whereas Whiston followed a cosmological argument based on Newton's discoveries and postulated the existence of a great comet spinning into the earth's orbital path and causing devastating floods.[27] Each in his own way attempted to explain the increase of water by natural—not supernatural—means.

These erudite and meticulously reasoned proposals were firmly based on evidence drawn from the various realms of natural philosophy: there were studies of fossils and sedimentary rocks, of rivers and the saltiness of oceans, meteorology, rain, earthquakes and mountain building, and a host of other topics that bore directly on the central theme of a universal deluge. With such intriguing material to investigate and marshal it is not perhaps surprising that the idea of a flood—a key feature in the strife-torn theologies of the later seventeenth century—should preoccupy scholars to the exclusion of other, possibly less fundamental components of the Noachian story. And indeed by the end of the century, the role of the Ark, once thought to be crucial, was distinctly commuted. There were now alternative issues, more critical problems to resolve, and other, more promising areas of natural history to explore. To the well-informed naturalist of 1691, to the acute and philosophical mind of John Ray as he despairingly wrote to his friend, the physician Tancred Robinson, Kircher's *Arca Noë* was nothing more than "trifling and superficial."

LINNAEUS AND THE EARLY HISTORY OF LIFE

By the time the young Carl Linnaeus (1707–1778) turned his attention to the world of nature, both natural history and the intellectual environment had changed to a marked degree. It is difficult in this respect to exaggerate the impact of growing numbers of species and natural history specimens known to European naturalists in the late seventeenth and early eighteenth centuries. The age of exploration, coupled with a continuous updating of existing animal and plant catalogues, had seen a virtual avalanche of species new to science. Where Ray described 150 quadrupeds, less than half a century later Linnaeus had to cope with nearly twice this number of mammals alone. Similar increases were recorded in every major class. During his lifetime Linnaeus listed the then staggering sum of 14,000 species, of which roughly 5,600 were animals. The period was patently one of collections, of natural history cabinets and private herbaria, and of the foundation of great national or institutional collections such as the Cabinet du Roi in Paris and the Musée in Saint Petersburg. Natural history had also benefitted from the contemporary enthusiasm for an unspoiled, primitive countryside, and collecting items from faraway places became all the rage. The very strangeness of the new material contributed to the growing realisation of an unbounded variety and richness in the remoter parts of the world. Besides such great collectors as George Clifford in Holland, Sir Hans Sloane in England, and René Antoine Ferchault de Reaumur in France, there were also many who collected on a smaller scale in their homelands or on their travels through continental Europe.[28] Naturalists were bombarded with new species from all sides.

And Linnaeus stood also on the threshold of the Enlightenment. Liberal ideas that had first emerged in the Renaissance were now becoming very generally accepted and, although religious authority was by no means disregarded, the Bible had come to occupy a more problematic position in explanations for natural phenomena. Many devout Christians continued to believe in the literal truth of the Scriptures, but others, like Linnaeus, were more heterodox, especially in their interpretations of nature. Linnaeus was, nonetheless, convinced of the reality and truth of the Christian religion. He had been destined for the Church before training in the medical sciences and saw himself consistently as the pastor of nature. Natural history was perhaps an ideal subject for him to pursue, for his deep Protestant faith could be fully expressed through a careful study of the Creator's work in nature. It was through the natural world that God spoke most clearly to man; it was man's duty and delight to investigate natural productions in order to reach some fuller understanding of His laws. Linnaeus considered himself "the publisher and interpreter of the wisdom of God," as the editor or translator of the organic text with which God had provided the human race. "This world," he emphasised in a notable preface on the meaning of natural history studies, "is destined to the celebration of the Creator's glory."[29] It would seem that nature, supplementing the authoritative word of the Bible, was Linnaeus's pathway to God.

It is easy to understand, then, the reasons which drew Linnaeus into a search for order and regularity in the natural world. Aptly enough, he considered the universe a gigantic museum collection given to him by God to describe and catalogue into a methodical framework. Linnaeus's classification schemes, his definitions, his rules for procedure and the binomial system of scientific nomenclature were formulated as direct responses to a great number of unknown species and his wish to provide a coherent psalter for the worship of their creator.

It is easy to understand also how the rapid escalation in the number of species affected Linnaeus's views on the early history of life and the story of Noah's Ark. Toward the beginning of the eighteenth century, the idea that not all animals could have been passengers on the Ark was becoming impossible to ignore. New forms of life were being discovered every day. Even if there had been a universal flood as Ray and others argued, the Ark must have been too small to hold all the species known to science. Linnaeus had little choice but to confront the dilemma that the previous generation had managed to avoid.

He entered the arena on the side of the angels, so to speak, and opted for a revised version of Genesis. Telescoping the story of Creation into that of the Deluge, Linnaeus presented a lucid biblical history capable of a literal interpretation and yet compatible with his personal experience of the natural world. He proposed that all living beings, including mankind, had their origin on a high mountain at the time primeval waters were beginning to subside. There could be no other explanation if one accepted the unmistakable evidence of a once universal ocean. Drawing on his knowledge of the recent elevation of the coasts of Sweden, backed up by reference to beds of shells now far removed from the sea, Linnaeus claimed in 1744 that dry land was ever increasing at the expense of the waters. Extrapolating backwards, he concluded that in the beginning only one small island had been raised above the surface of a worldwide sea. This promontory must have been the site of Paradise and the first home of living beings.[30] Echoing Sir Matthew Hale's argument, he further claimed that every species began with only two members, so making it probable that parts of the story of the Garden of Eden were true and that animals had, for a time, shared the same locality:

> If we therefore enquire into the original appearance of the earth, we shall find reason to conclude, that instead of the present wide-extended regions, one small island only was in the beginning raised above the surface of the waters. If we trace back the multiplication of all plants and animals . . . we must stop at one original pair of each species. There must therefore have been in this island a kind of living museum, so furnished with plants and animals, that nothing was wanting of all the present produce of the earth.[31]

True to his principles of order and regularity, Linnaeus decided that this living museum could not be supposed to have existed as one chaotic collection tumbled together on a metaphorical workbench. Taking an example from his botanical predecessor, J. P. de Tournefort, he suggested that the Paradisiacal mountain presented a wide range of ecological conditions, arranged as belts from tropical through temperate to polar zones. If they are sufficiently high, he argued, mountains of the present day show exactly this kind of climatic stratification. It is perhaps significant that Linnaeus took Mount Ararat as a suitably authoritative, modern example of successive zones of temperature. Although the foothills were rooted in the deserts of the plain, Ararat undeniably carried permanent snow at its summit; it represented in summary the various climatic zones that circle the northern hemisphere. Linnaeus transferred this observation to his single aboriginal mountain, postulating that each pair of animals was created in a particular climatic belt in association with the other species designed for such conditions. From reindeer and arctic lichens at the summit to tropical palms and monkeys at the foot, every plant found its "proper soil" and every animal its "proper climate." The mountain was an entire world in miniature.

Then, as the primeval waters receded, Linnaeus envisaged organisms moving in turn to the latitudes and environments where they were to remain for the rest of time. Presumably it was a combination of sheer chance, the speed of migration, and the helping hand of God that determined exactly where the animals and plants ended up. Much of Linnaeus's essay on the "Increase of the habitable earth" was given over to a thorough discussion of the multifarious means of dispersal that organisms might have used for their journey.

But the Ark played no part in Linnaeus's story. "Is it credible," he asked in 1744, "that the Deity should have replenished the whole earth with animals to destroy them all in a little time by a flood, except a pair of each species preserved in the Ark?" To him, this was as absurd as believing that the Deluge came almost without warning and went as suddenly as it came. Rational thinkers, he stressed, would reject these suggestions as being contrary to either reason or practical experience.[32]

The story of Noah's Ark was thus scrutinised and found to be wanting. Linnaeus could not conceive how all the animals of the world—including the 5,600 species he had personally named and catalogued—were to be preserved in a wooden barque and then supposed to repopulate the earth for a second time. Yet it should be emphasised that Linnaeus was not prepared to relinquish other aspects of the legend. He saw no reason to contest the biblical account of creation, and he retained the idea of a Noachian deluge in the form of a once universal ocean. The significant point is that the Ark was the first component of the story to be questioned. Evidently it had outlived its usefulness as a guide to studies of the primitive earth. Nevertheless, Linnaeus recognised the story's overall value as an explanatory model, and incorporated the essence of the legend into his own more secular account; he too had a mountain, a vast expanse of water, and a set of animal pairs migrating from a single geographical centre. Linnaeus had the best of the Ark story without the intellectual impediment of the Ark itself. The key to this interesting mixture of ideas lies, it would seem, in his concept of species.

THE SPECIES CONCEPT

Linnaeus's definition of species and his desire to classify the living world were, of course, strongly influenced by the issues of his day. The species problem, always of interest to naturalists, was then coming under renewed attention as the rigid classification schemes of earlier philosophers gradually lost ground to those with a more flexible and natural arrangement of classes. What, they asked themselves, *was* a species? The problem had also been exacerbated by the sudden influx of strange specimens from various parts of the world, not all of which could be fitted into current systems of categorisation. What, if anything, was the meaning of such diversity?[33]

So Linnaeus, as others before him, searched for the fundamental quality, the quintessential property of the living world that would define and characterise the phenomena he could see around him. This he found in the idea of "species." To him species were discrete and indissoluble units, inextricably binding individuals together by descent, generation after generation, in some constantly determined way. Not simply a collection of similar beings, Linnaeus's species were metaphysical constructs that proved by their very existence that nature was at root capable of an ordered arrangement, an arrangement which had been, and was to remain, always the same.

In a slim pamphlet published in 1735, the *Systema Naturae*, he set out the propositions that had led him to such conclusions. The argument ran on lines similar to those proposed by Matthew Hale about sixty years before. Every individual, Linnaeus pointed out, is propagated by an egg and every egg produces an offspring closely resembling the parent: "Hence no new species are produced at the present time." Since individuals are also capable of adding more than one offspring to the next generation, the number of members in each group must be growing: "The number of individuals in each species is greater now than it was at first." If we count backwards from the present, continued Linnaeus, it is evident that the series ends in one set of parents, or in a single hermaphrodite being. The origin of these aboriginal parents must be attributed to "some omnipotent and omniscient Being, namely God, whose work is called Creation."[34]

Since species were only species by descent, Linnaeus credited God with producing the first individuals. He asserted that there was a real living bond between the first and the most recent organisms of the world and insisted that every animal of the present day was descended from the original pairs that had been created soon after the dawn of the habitable earth. He supposed an unbroken line of living beings stretching from the Garden of Eden to the present day, arguing that "all living things, plants, animals, and even mankind themselves, form one 'Chain of Universal Being,' from the beginning to the end of the world: in this sense truly may it be said, that there is nothing new under the sun."[35]

Of course Linnaeus was not the first to have this thought. Everyone who had written on the story of Noah's Ark had held exactly the same view. But he did draw the idea of a creation much more closely into current scientific thinking. By linking the first representative of a species to the latest, Linnaeus made the study of origins a legitimate subject, and by placing great emphasis on the fact that it must have been the Creator who gave each living being its shape and constitution he showed the way to investigate these origins. Although species were still held to be a product of divine will, and the moment of creation still a mystery, it was clear that naturalists could study animals and plants confident in the knowledge that they were precise replicas of the first living beings—replicas of the organisms fashioned by the Lord himself, in the shape that He thought best. The Linnean approach consequently enhanced the importance of the adaptation and suitability of individuals to their surroundings and revitalised inquiries into the relation between form and function.

Although the species concept primarily focussed scholarly attention on God's plan for nature, more important for the history of biogeography was the parallel idea that every form was suited to its function. Under Linnaeus's scheme of things, plants and animals had to be seen as perfectly constructed for the environment in which they were situated for the rest of time. The whole system was impeccably and delicately balanced; every organism had its place and was an essential link in a complex, self-regulating cycle of existence. Every living thing was both predator and prey. Insects prevented plants from dominating the world in the same way that birds kept insects within reasonable bounds. Birds, in turn, were regulated by carnivorous quadrupeds of one kind or another. Each animal and plant species had a place to fill and a task to execute. Linnaeus assumed these functions were fulfilled through a hidden struggle in nature, where every species produced more offspring than could possibly survive. Such overpopulation generated intense competition between organisms and maintained the exquisite equilibrium of the whole.[36]

First worked out in his *Oeconomia Naturae* (1749) and the *Politia Naturae* (1760), Linnaeus's views on the harmony of nature were of great importance. The reciprocal action of competition on the one hand and high reproductive rates on the other produced a perfectly balanced population. Yet despite the vibrant activity, the birth and death of countless individuals, that seems to underlie Linnaeus's hypothesis, the net result was one of stasis. He saw organic nature as a static picture, with every tiny detail preordained by the first creator. God was the author of these complex, interrelated systems because He had given species their form and decreed their function, each organism being designed for its particular role in nature.

Linnaeus's stress on "places" and the suitability of every species for its particular station and function in the drama of existence required careful consideration of the origin of these complex relationships. He decided that even the very first species must have lived in some sort of defined association, in environments that were precursors of later conditions; otherwise, how would they have survived? So his aboriginal mountain was not a literal Garden of Eden where animals of all varieties mingled freely, but a strictly arranged succession of climates and populations, where reindeer, for example, were confined to the coolest regions of the primeval land and palms to the warmest. Once the habitable earth had been revealed, and animals and plants had spread out into the virgin territories of the globe, the process of populating the earth was over and could not be repeated. If there had been a deluge after this, a deluge that destroyed some species and an Ark that preserved others, the original God-given equilibrium and stability of nature would have been destroyed. Such a state of affairs was inconceivable to the young Linnaeus. For him, nothing had been added to or taken away from the animal kingdom since the first days of creation. (Later, as is well known, Linnaeus changed his mind about the absolute stability of nature and decided that plant and animal genera must have interbred, or hybridised, after the first creation.) To allow the reality of the Noachian story would overturn all that he had discovered about the fixity of species, the economy of nature, and the direct relationships between the organisms of the beginning and those of his own time. The latter concepts in his opinion were more essential than a belief in the literal truth of the Bible. Noah's Ark was abandoned.

THE ABANDONED ARK

Since species had been created, according to Linnaeus, in the shape they now possessed, and this shape was the most appropriate one for the conditions they lived in, then it must follow that species had also been created in the station they now occupied. "Form," "suitability," and "place" were now an indissoluble threesome. With Linnaeus's work there came together the notion of different types of animals and plants and the notion of the different places or spaces occupied by them. In other words, since species were real entities connected by descent to a distant ancestral pair, they must have been created *like* today's forms and under conditions identical to those *where* they are now. Linnaeus himself went to some trouble to ensure that this duality was not contravened in his account of the creation process.

Unfortunately for his argument, however, Linnaeus's thinking opened the door for more secular accounts of geographical distribution. By arguing so effectively for a close link between the structure of an animal and its physical surroundings, he unwittingly endorsed a rival theory of distribution first proposed by Buffon and taken up by Johann Reinhold Forster, Eberhardt Zimmermann, and Karl Willdenow, among others. Each Linnean species was so well suited to its native habitat that scholars could not imagine it travelling across half the world from Syria. The reindeer, it was argued, would have to pass over some of the most difficult and inhospitable lands in the world to get from Ararat to Lapland. If it was designed for the cold, how had it survived in the deserts en route? Perfect adaptation of organisms to their place in nature, as dictated by Linnaeus, simply was not compatible with migration from a single source. It was more feasible to inject new life into the old theory of La Peyrère and to assert that species had been created in many different places, each animal in the country where it was destined to live.

The Comte de Buffon (1707–1788) was, perhaps, the most famous exponent of this revitalised hypothesis. His philosophy was remarkable in his own time for its comparative freedom from traditional thought and religious scruples. Diametrically opposed to Linnaeus and often completely misunderstanding his intentions, seldom taking the trouble to describe and precisely sift through the multitude of facts at his disposal, Buffon nevertheless managed to formulate thoughts and theories of breathtaking insight. In his great *Histoire Naturelle, Générale et Particulière* (1749–1804) he recognised that for practical reasons each geographical region must possess different forms of life. In Buffon's view, the most striking fact of distribution was the combination of differences and similarities between the populations of the New World and the Old; herein lay the answer to the question whether all animals had indeed originated in a single spot.

First of all, asserted Buffon in 1753, it must be recognised that there were animals living in the Americas that were wholly endemic to that region. The puma and the jaguar were not, in his view, misplaced lions or tigers, but truly different species of *Felidae*. The forms he held to be native to each of the "worlds" were typified by being occupants of the "torrid zones" and incapable of maintaining or reproducing their kind in colder areas. Topography and climate cut them off, being confined on two or three sides by seas they could not cross and on the other by lands too cold to support them. "Not one of the animals of the torrid zone of one continent," he wrote, "are natives of the torrid zone of the other."[37]

Their differences could be explained by a combined geologicoclimatic hypothesis. Buffon seems to have believed that all organic life originated near the north pole in some warmer period and had moved south over the globe as the earth progressively cooled. Because the three major landmasses of the world were juxtaposed at their northern extremities and separated by vast oceans in their southerly parts, this flow of organisms from the north to the south meant that the original population was rendered disjunct. Once the northern areas of the landmasses cooled down, they effectively isolated each set of species in a topographical and climatic cul-de-sac.

In the New World this was followed, Buffon thought, by some kind of structural degeneration from the primary type caused by the debilitating environment. "All animated nature is comparatively diminutive in the new continent," he stated; "Something in the combination of the elements and other physical causes opposes the aggrandisement of animated nature."[38] But he was hard pressed to identify the causes of this diminution. He attempted to link what he imagined was the small stature and reduced vigour of American species with a corresponding decline in the manners and social structure of the country's human inhabitants. Native Americans, he reasoned, had little cultural cohesion and no interest in taming or domesticating nature, a quality which should have led inevitably to a "strengthening" of the region's plants and animals and to a greater overall diversity. But however this diminution came about, there could not have been any communication between tropical Africa, Asia, and America. Should the species of these regions be different—as Buffon claimed they were—it was evident that they must have originated *in situ*. By logical inference, then, they had not migrated there from the Biblelands.

Yet Buffon did not abandon the concept of migration entirely. Although he rejected it as an explanation of the common origin of *all* species, he was quite prepared to use it to explain the evident similarities of certain organisms living on either side of the Atlantic. Animals such as the Old World ox and the American bison were patently alike, and Buffon supposed that they would turn out to be one and the same species—the test being whether they were able to produce a fertile hybrid offspring. The closeness of their physical structure encouraged him to think that the Eurasian form had migrated at some stage from the Old to the New World, where it had degenerated slightly according to the immediate demands of its surroundings. Oxen may have found a passage across northern districts before the land became too ice-bound to sustain vegetation and, "after having advanced into the temperate regions of this New World, they received the impressions of the climate, and in time became Bisons."[39]

One had to judge the animal life of the earth according to the climate and the geographical "disposition" of each region, he continued, bearing in mind that the earth had once been very much warmer than today. These considerations led him to assert as a "general rule" that the mammals of the tropical regions of the New and Old Worlds were exclusively confined to their own areas. So far as it went, this was an explicit statement of the origination of sets of species in the areas they currently inhabited.[40] Buffon had no use for either the story of Noah's Ark or the idea that all animals had a common source in Armenia. Scarcely ten years after Linnaeus's short papers on geographical distribution and the early history of the earth, Buffon presented an entirely different and emphatically secular account of the same phenomena.

The *coup de grâce* for the Ark came twenty years later in an interesting and little-known zoological tract by Eberhardt Zimmermann, *Specimen Zoologiae Geographicae Quadrupedum* (1777), the first book to describe in detail the distribution of mammals. Zimmermann ridiculed the idea that only one pair of each kind of animal was created. On this reasoning, coupled with Linnaeus's living museum, the first pair of lions would eat the first pair of sheep, soon followed by the first pair of goats and other herbivorous species in quick succession. Finally, even the lion would die from hunger after it had eaten everything in sight. No, wrote Zimmermann, it was more rational to believe every animal was created in the area where it now lives, under the same climate that it now enjoys, with the same food rations already in abundant supply. In effect, he argued that an entire ecosystem was called into existence in one fell swoop.[41]

With this idea established there was no further reason to continue believing in the Ark or in original pairs. Zimmermann has the distinction of being one of the first naturalists to hint at multiple creation; his work not only signalled the completion of a major transformation in the way scholars approached the Bible, and in the way they used the Mosaic account of creation, but also indicated the route that future investigations were to take. Once scholars had established beyond all doubt the impossibility of every living thing having a common source in Ararat, it then became conceivable to ask more basic questions about the creation of species. Once the concept of one geographical source for all species had been abandoned, the opportunity arose for the contemplation of other, more far-reaching philosophical issues. It was not the Ark that perplexed and fascinated post-Linnean naturalists, but the history and origin of species. By Zimmermann's time the Noachian story had been thoroughly adapted to the demands of the temporal scientific world—so much so that it was relatively easy for mid-eighteenth-century naturalists to abandon the whole account as legend. The Ark had more or less been reasoned out of existence.

It is usual to attribute the demise of belief in Noah's Ark to an inability on the part of thinking men to fit growing numbers of species inside it, to suppose that the pressure of numbers, the inescapable data of science, compelled naturalists to relinquish this favoured item of religious dogma. Certainly, the number of species known to scholars *did* grow rapidly in the three-quarters of a century between John Ray and Zimmermann, increasing from 150 quadrupeds in 1691 to over 450 mammals alone in 1777. But this is only part of the story.

The Scriptures presented a coherent, detailed, and explicit account of the early history of animal life. Nevertheless, it seemed impossible to justify any single aspect of it without calling others into doubt. The more scholars attempted to explain the story, the more difficult it was to believe in it. Without intending to do so, literalism successively destroyed facet after facet, so that on both philosophical and practical grounds the saga of Noah became increasingly irrelevant. Ultimately, authors became so accustomed to using the pragmatic evidence of their senses in natural philosophy that it seemed perfectly rational to dispense with the Ark altogether.

However, the rejection of Noah's Ark as an explanation for geographical distribution of the animal kingdom did not invalidate the results of pre-Linnean researches any more than it devalued the methods of inquiry that had developed around this particular model. The intellectual achievements of scholars working within the framework of the Bible provided later naturalists with essential tools for studying the multitude of unexpected problems revealed by the absence of the Ark. Noah may have disappeared, but he left an impressive display of empirical methods behind him.

FLORAS AND FAUNAS

Although they had ignored Linnaeus's explanation of geographic dispersal, his contemporaries nevertheless used the species concept to construct alternative schemes in which a single, biblical centre for all species was replaced by the idea of many centres of creation, each species in the area in which it now lived. This move was made possible by Linnaeus's demonstration of the close links between living things and their natural surroundings. By postulating that the Creator had "designed" each and every organism for its specific place in nature, the species concept implicitly joined together the notion of organic "kind" with that of topographical "place."

This view was not exclusively concerned with single species and their relations with the environment. It was also instrumental in encouraging naturalists to think of groups of species, of a set of organisms intimately connected to each other and to their own geographic area. With such ideas there came a gradual formulation of the modern concept of "floras" and "faunas," a concept that conveys not only a simple list of the different forms of life existing in any area, but also a sense of the structural uniqueness and topographical limits of the entire population. Floras and faunas came to be seen as coherent units. The advent of this important notion finally closed the door on theories of distribution that relied on biblical literalism, for built into the idea of a flora or a fauna was the assumption that each set of species was, in some way, native to its country. Floras and faunas could not have a common origin in Mount Ararat, nor could their individual members have migrated piecemeal from Armenia to reassemble at their ultimate destination. Having swung away from the notion of one source for all species, naturalists turned instead to the marked regionality of the living world.

Linnaeus's pupils were among the first to capitalise on these new, particularly eighteenth-century ideas. They were sent far and wide, rejoicing in the title of "apostles," to gather information about regional floras. Pehr Kalm travelled through North America during the years 1748 to 1751 and Pehr Osbeck sailed on East Indian merchant ships to China. Daniel Solander and Anders Spaarmann accompanied Captain Cook across the Pacific on his first and second voyages, respectively. Linnaeus arranged to publish the works of his pupils Hasselquist and Loefling, after their journeys through the Middle East and central Spain: *Iter Palaestinium* came out in 1757, closely followed by his friend's *Iter Hispanicum* in 1758. One enterprising student, Carl Peter Thunberg, managed to gain entry into Imperial Japan, an achievement duly recorded in his classic *Flora Japonica* of 1784. Johann Gmelin and Rolander travelled to Siberia and Surinam, returning in triumph with collections to enhance the Linnean herbarium. Several apostles, on the other hand, did not come back at all, causing the paternalistic Linnaeus profound grief and guilt. But the students continued to travel: Falck (in Tartary), Mutis (in Colombia), Koenig (in Tranquebar), and Forsskal (in Arabia) survived to tell their tales, and they forwarded many seeds and plants to Uppsala during the last third of the century.

The work of these disciples reflected the new trend in natural history. Up to the middle years of the eighteenth century, the mass of information about different species in different places had still not been resolved into the idea that certain areas carried particular floras and faunas. North America was the home of Walter Raleigh's potato, yet there was little notion of an American flora as such. Rhododendrons and azaleas were known to live in Japan and parts of continental Asia, but there was no clear idea of the group as a typical and characteristic inhabitant of the Indo-Asiatic landmass. It is true that the predecessors of Linnaeus included all the animals of a particular region, or all the plants, under the rubric "fauna" and "flora," but the relationships between these organisms and the areas where they lived was not concretely recognised. Instead, the terms were used more as labels for lists—as descriptive catalogues of the organisms that had come to the attention of workers in the field. Such lists were not meant to be exhaustive surveys of the life contained in any specific area, and the words "fauna" and "flora" were merely synonyms for expressions such as "historia animalium" or "catalogus" or for the term "florilegia." One of the earliest known examples of this particular usage came in the title of Simon Paulli's book *Flora Danica* (1647), closely followed by John Ray's *Flora, seu de Florum Cultura, Or, A Complete Florilege* in 1665.

By the late 1760s, however, a gradual transformation in the way that groups of animals and plants were considered had taken place. Now they were seen to be discrete units, each one special to its own region of the globe. The meaning of the words "flora" and "fauna" appears to have changed accordingly, especially in the titles given to publications by Linnaeus's followers. Here, "flora" and "fauna" were supposed to convey a certain comprehensiveness, a particular geographic region that had been thoroughly investigated and exhaustively catalogued. To write a Linnean-style *Flora* or *Fauna* was to describe a self-contained and precisely located assemblage of organisms. It would be misleading, perhaps, to place too much emphasis on a shift in the meaning of these terms, since the earlier usage (as in catalogue or list) is still valid today. Also, the titles of natural history works are often too brief or allusive to reveal the full intentions of the author. Moreover, the historian's difficulties are exacerbated by the fact that many of the areas described in *Flora*s and *Fauna*s are not, of themselves, distinct in any natural history sense: rather, they often are (or were) purely geographical entities, like islands, or regions circumscribed by solely political boundaries. Nevertheless, it was still possible for botanists to describe the flora of a political state, such as Sweden, or an island, such as Britain, as if it were a natural floristic unit, because Sweden and Britain were parts of other, genuinely self-contained botanic regions (Scandinavia and northwest Europe). So it does seem to be the case that, after about 1760 or so, the words "Flora" and "Fauna" began to appear more frequently on natural history title pages, and always in conjunction with a text that dealt with discrete groups of species.

J. F. Gronovius, one of Linnaeus's closer friends, started the fashion, if we may call it that, with his *Flora Virginica* (1739–1743). Linnaeus published his own *Flora Suecica* in 1745 and the *Fauna Suecica* in the following year. His *Flora Zeylandica* appeared in 1747. One of the apostles, J. G. Gmelin, came out with a *Flora Sibirica* in 1747 (1747–1769), which he followed with a *Flora Orientalis* in 1755. In 1760 John Hill's *Flora Britanica* [*sic*] was published, outstanding for being the first Linnean Flora of Great Britain. Antoine Gouan produced the *Flora Monspeliaca* in 1765, followed by yet another of Linnaeus's disciples, G. C. Oder, writing on the *Flora Danicae* (1766–1799). Johann Reinhold Forster not only wrote his own *Flora Americae Septentrionalis* in 1771, but also translated Pehr Kalm's work into English in the same year. An Italian, Giovanni Scopoli, published the *Flora Carniolica* (1772) in Vienna, a city that later produced one of the more sumptuous Floras, N. J. von Jacquin's *Flora Austriacae* (1773–1778). From Britain came the *English Flora* by Richard Weston (1775), and two in the same year, William Curtis on the *Flora Londinensis* (1777–1798) and John Lightfoot on the *Flora Scotica* (1777). In France, Lamarck presented his only botanical tract, the *Flore Française* in 1778.

The term "fauna" seems to have taken longer to become popular, perhaps because there were so many well-known alternatives, such as the "zoology" of a region, or because the restricted groupings of birds, mammals, insects, and suchlike lent themselves more readily to monographs than to the study of an entire faunistic area. Looking at it another way, modern zoologists would probably claim that regional groupings of animals have always been more difficult to discern than those of plants. It was not really until the nineteenth century that "fauna" appeared with any regularity in the titles of zoological works, although there were at least three notable exceptions: Otto Fabricius on the *Fauna Groenlandica* (1780); Scopoli on the *Deliciae Florae et Faunae Insubricae* (1786–1788); and G. W. F. Panzer on the *Fauna Insectorum Germanicae* (1793–1813). In 1800 Retzius produced an amended version of Linnaeus's *Fauna Suecica*, followed by J. C. Mikan on the fauna of Brazil (1820) and by Sir John Richardson with his *Fauna Boreali-Americana* of 1829–1837. J. Holandre wrote on the *Faune du department de la Moselle* in 1836, Jonathan Couch on *A Cornish Fauna* in 1838, and Edmond Selys-Longchamps on the *Fauna Belge* in 1842. Minor works by the authors Alfred Malherbe (*Faune Ornithologique de la Sicile*) and Eduard Eversmann (*Fauna Uralensis Lepidopterologica Volgo-Casanae*) followed in quick succession in 1843 and 1844, and were joined by those of more noted zoologists when Hugh Falconer published his *Fauna Antiqua Sivalensis* (1845–1849), Philipp Franz van Siebold the *Fauna Japonica* (1823–1850), and Moritz Richard Schomburgk the *Versuch einer Fauna et Flora von Britisch-Guiana* of 1848.

In the same spirit, though not actually using the word "fauna," Thomas Pennant described a strictly regional association of animals in his *British Zoology* (1761–1766). Although he did not at first bother with Linnaeus's views on classification, Pennant later came to catalogue other faunas in the binomial manner. His *Indian Zoology* (1769) and the *Arctic Zoology* (1784–1785) exhibited a keen sense of the integrity of local animal populations. The same awareness is to be found in letters written to the Honorable Daines Barrington by Gilbert White, rector of Selborne. White referred to faunas and floras in this expanded, post-Linnean sense, telling Barrington that he expected Linnaeus, of all people, to be able to "account for the motions and manner of life of the animals of his own Fauna," and noting that "Chalks, clays, sands, sheep-walks and downs, bogs, heaths, woodlands, and champaign fields, cannot but furnish an ample Flora."[42]

These verbal distinctions reflected a corresponding change in the way natural historians approached the topic of geographical distribution. In the last third of the eighteenth century, the arrangement of the living world was seen as a problem that involved many different sets of species, each arranged into a group. The anatomical distinctions and the topographic limits of each association had to be investigated and described, and individual organisms had to be understood as part of a much larger whole. Each flora and fauna was thought of as a unique living entity characteristic of one geographical region alone.

Here, in this assumption, lay the roots of the major theoretical achievement of post-Linnean distribution studies. Here lay the foundations of the concept that was, in one way or another, to dominate ideas about animal and plant geography in the late eighteenth and early nineteenth centuries. Although it was not to become manifest until the works of naturalist-philosophers like J. R. Forster and Alexander von Humboldt, the idea of "biological provinces" had begun to capture the scholarly imagination.

TWO

The Nations of Plants

Virtually synonymous with the idea of regional floras and faunas, the concept of biological provinces was in essence simple. The notion of an integrated community of organisms was linked to the more obvious idea of a purely geographic or physiographic region and led to the recognition of sets of species intimately connected to each other and to their own geographical area. It was a synthesis of the living and nonliving subdivisions of the earth's surface.

This important idea was at the heart of late-eighteenth-century distribution studies and proved equally crucial in later years. By investigating the qualities of biological provinces—their boundaries, their member species, their physical attributes and topographical features—scholars could pursue their subject with a clearer focus. More significant for the history of natural history, scholars could now approach the issue that had been left so naked by the final disintegration of the Noachian story: that is to say, they could proceed to speculate about the origin and history of species. If animals and plants did not come from Ararat, where *had* they originated? Geographical distribution and the investigation of the world's major biological associations promised to reveal at least some of the answers to this weighty question.

Changing views about the scope and meaning of geographical studies brought changes of another kind, no less interesting to the historian. During the later eighteenth century and in the beginning of the nineteenth, the majority of researches in this subject were concerned with the kingdom of plants in preference to that of animals. Although animal distribution had long been the centre of scientific attention, it now seemed to lose its relevance and to give way to phytogeography—a topic scarcely investigated before this time. A newfound interest in plants united students of distribution, just as a newfound interest in provinces drew the theorists of the subject together. The study of regional groups of plants suddenly came into its own.

Much of this change in emphasis must be due to the demise of Noah's Ark as a feasible explanation for zoogeographic patterns, to the growth of a belief that species are native to the areas in which they are found. The story of the Ark, of course, had little to offer naturalists interested in the history of plants beyond the suggestion that useful cereals and certain selected species (such as the grape vine) had been transported by Noah. The legend of the olive branch indicated that plants and trees were somehow unaffected by the deluge, or at least had a history different from that of animals. Moreover, the Noachian story stressed the act of migration above all else, so it seems reasonable that the earliest commentators should have concentrated almost exclusively on organisms with obvious locomotive powers: the quadrupeds and birds. When Noah's Ark was replaced by the idea of many creations, each species originating in the appropriate area, scholars accordingly shifted their gaze onto living beings that do not move around, namely trees and other plants. Plants, much more than animals, exemplified new ideas inherent in the study of biological provinces and the subdivisions of the earth's surface: plants are (literally) rooted in one particular region; their life-support systems are necessarily tied to the environment in which they live; and they form distinctive vegetations, such as forest and savanna, that are in turn linked to specific climatic and topographic features of the earth's crust. Also, as it must have seemed at the time, plants were more thoroughly catalogued, more readily identified, than animals. Botany had well-developed scientific resources to offer its practitioners, including the accumulated literature of centuries of taxonomic research, the practical expertise of travelling collectors handed down through the generations, and the great herbaria of Europe available for reference. Here, then, were the ideal subjects for an inquiry into the natural "regions" of the organic world.

NATURAL REGIONS

One of the most significant late-eighteenth-century studies of geographical distribution was conducted by Johann Reinhold Forster (1729–1798), the German philosopher-naturalist who tried to synthesise Buffonian and Linnean concepts into a single unitary scheme. Arriving in England in 1766 in the hope of making his name and fortune as a scholar, Forster for some years led a hand-to-mouth existence alleviated only by the publication in 1770 of his Linnean-style *Catalogue of British Insects*. Eventually he became a lecturer at the Dissenters' Academy in Warrington, Lancashire. There, he was chosen to accompany Captain Cook on his second expedition to the South Seas, replacing Sir Joseph Banks at the last moment. It was the golden opportunity for which Forster had longed.

Cook's second tour of the southern hemisphere (1772–1775) produced more in published results than the first, but was not as socially or professionally rewarding as Johann Forster had hoped. Forster and his son Georg (who accompanied him on the voyage) were both religious and straitlaced, and did not get on well with the sailors. Indeed, prolonged exposure to the two Forsters persuaded Cook never to allow a naturalist aboard his ship again. Nor were things any better after the voyage. In 1775, when H.M.S. *Resolution* and *Adventure* returned home, having circumnavigated the globe and claimed Australia for the Crown, Cook persuaded the Admiralty to forbid Forster to publish any account of his discoveries. But the ban did not apply to Georg Forster, who borrowed his father's notes to write his own book, *A Voyage Round the World*, in 1777. It was only after an extended controversy with the Admiralty—during which Cook published his *Voyage towards the South Pole, and Round the World*—that the elder Forster managed to get his book into print. Entitled *Observations made during a Voyage round the World* (1778), it was a studiously factual and scholarly work.

It is in this ill-starred volume that Johann Forster's statements on geographical distribution are to be found. He approached the study of the natural world as a unified and unifying whole and tried to interlace the physical geology, geography, and climate of a country with its vegetation, animal life, and human occupants, including for good measure its agricultural practices, local manufactures, and the customs or "manners" of the populace. He combined social comment, politics, plants, and anthropology into a coherent assessment of each region visited. Every area differed according to its position on the globe, and Forster saw his job as the comparison and contrasting of these discrete geographical entities. The work was rooted in a concept of regional integrity, in the belief that organisms occur naturally in sets, as it were, each special to one location on the globe.

Plants, in particular, showed Forster the general topographic and climatic characteristics of each country: the spread of groupings such as Asiatic rain forest reflected the range of a specific combination of physical properties, incorporating rainfall, temperature, bedrock, distance from the sea, and so on. When these inorganic features changed in quantity or quality, the plant life changed too. In this sense, vegetation was the key to comprehending mysterious regularities and patterns in the natural world. Forster therefore took pains to show how floras replaced each other as the physical characteristics of the earth's surface altered, each unit gradually giving way to the next in an orderly sequence. In the Pacific islands, he noted, the plant life of the New World met and mingled with that of the Old. Each flora extended to the east or west, gradually becoming less like its parent and more like each other as the distance between the two decreased: "The easternmost isles [of the Pacific] contain a greater number of American, than of Indian plants; and again, as we advance further to the west the resemblance with India becomes more strongly discernable," he explained.[1] It was inconceivable that the physical features of the earth did not alter over this enormous section of its crust, and Forster therefore assumed that the floristic characters would also alter accordingly. One giant botanical unit was transformed into another.

Like his contemporary Eberhardt Zimmermann, Forster recognised that those species common to islands and continents decreased in inverse proportion to their distance. He thought this was only to be expected, because islands were increasingly detached from the set of physical features that characterised the continents concerned. For one thing, the climate of islands was always more equable than that of continental landmasses. For another, oceanic islands were frequently volcanic in origin and therefore composed mostly of rocks unlike those of larger lands. These physical peculiarities precluded any absolute identity of species. Furthermore, it was just such inorganic peculiarities that explained why islands possessed proportionally fewer species than their continental counterparts. The environment, suggested Forster, regulated the organic world to such an extent that even the number of species in an area was proportionate to available physical resources: "Islands," he wrote, "only produce a greater or lesser number of species, as their circumference is more or less extensive."[2]

Forster's evident appreciation of plant life led him beyond these points. He recognised the important role that vegetation had to play in determining the kind of animals found in a particular region. Human colonisation—and civilisation—also depended on the plant life of any given country. Under the heading "Stations," he went on to describe how the races of man in the Pacific area were related to one another and to their supposed sources in neighbouring landmasses. Just as the plant life of the region changed over the sweep of the Pacific, Forster believed he could discern a gradual transformation in the "types" of human being, and he concluded one giant human "province" must therefore be giving way to another. Most of the people of the South Sea Islands belonged to one racial group, which in his opinion had originally emigrated from the Asian continent; these peoples were completely different from the darker Negroid inhabitants of New Guinea and the islands close to the Australian mainland. Following Buffon's anthropology, Forster attributed the minor differences between the occupants of one island and the next to the debilitating effect of the environment. This study was entirely analogous to his work on plants.

Taking a broader view, Forster applied these general ideas to the whole world: the distribution patterns of plants, of animals, and of men were tied together into one grand scheme linked to the surface heat of the earth. During the sea voyage he had been struck by a more or less uniform decrease in the richness and diversity of floras from the equator to the poles. This correlation—one which other botanists had certainly noticed, and which Buffon had clearly pointed out—was extended by Forster to include animals and men. There was, he believed, a general loss in "splendour," "nobility," and variety in countries far from the tropics:

> The animal world, from being beautiful, rich, enchanting in the tropics, falls into deformity, poverty and disgustfulness [*sic*] in the Southern coasts. We cannot help being in raptures, when we tread the paths of O-Taheitan groves, which at each step strikes us with the most simple, and at the same time the most beautiful prospects of rural life. . . . Let us remove from hence to the temperate zone: what a falling off from the soft scenes of domestic quiet, to the wilds of New Zealand! The rocky mountains, the forests, yea, human nature itself, all look savage and forbidding. . . . But as we advance to the South and cross an immense ocean, in the midst of which we see some lonely birds, skimming the waves and collecting an uncertain sustenance, we arrive at the South end of America, and view a barren coast, inhabited by the last and most miserable of men, and but scantily clad with low and crooked shrubs.[3]

For Forster, nature appeared to reflect the distribution of the surface heat of the earth. Like Buffon, he believed in the theory of a central heat, which stipulated that the planet had progressively cooled from its incandescent state to the present unequal distribution of heat in different latitudes. The two men also shared the romantic view that the element heat had some esoteric capacity to stimulate organic variety.

This emphasis on heat as the ultimate reason for the appearance of nature was significant in several ways. It meant that diverse facts of distribution and seemingly disparate floras and faunas could be interpreted as the various manifestations of a single causal factor. For Forster, the organic world could be arranged in two huge spectra originating at the equator and finishing at the poles, mimicking the temperature belts of the earth. These two "living" hemispheres more or less reflected each other in the diminution of the "strength" and variety of natural resources as distance from the equator increased. Forster's belief in such a spectrum unified and gave validity to his distribution studies and encouraged him to think of the subject as a global affair dependent on one basic explanation: heat.

Moreover, Forster's preoccupation with heat led him to the proposal that the tropics were the source of life—or if not the source, the area in which nature reached its highest and most diversified expression. This theory stood not only for plants and animals, but for human beings as well. Because they lived in the heart of the tropics, where food was plentiful and manners were simple, he thought the Taheitans had the highest form of civilisation. In contrast, the inhabitants of Tierra del Fuego were as savage as their environment was miserable. Forster's book was a paean to the Enlightenment vision of nature more usually associated with Rousseau; he glorified the natural behaviour, the "rural life," and the "domestic peace" of the South Sea Islanders, and compared these unfavourably with the harder life of temperate and boreal zones. There was no doubt in Forster's mind that the warmth of the tropics was the source of all the good and admirable things in the world: the provider of food, of unsophisticated human society, and of natural beauty. In short, he linked the physical world with that of human society and of morals. And by insisting that plants acted as the mediator between these two worlds, Forster elevated botanical geography to a hitherto unrecognised position of importance.

Forster evidently possessed a refined sense of the uniqueness of botanical and other populations, and of the links between these and their geographical location. There was, in his opinion, a close relationship between the organic and inorganic elements of the earth's surface. His work of 1778 marked the close of a period, little more than fifty years long, in which natural historians came to recognise that points such as these were not trivial but concepts of far-reaching significance. Forster, more than any of his predecessors, understood floras and faunas as living communities that ought to be considered as a whole. He emphasised that these units were characteristic of a single geographical area, which ought therefore to be taken as its natural home. Furthermore, an increasing sense of the intricate relationships between species and their physical environment led him to reject a common point of origin for all forms of life in favour of a scheme stressing the existence of many, geographically diverse sites of creation. Most important, he asked himself why this should be so, and decided that heat was the fundamental factor. He had, as it were, set the stage for more sophisticated enquiries into the geographical and historical relations between groups of living beings. He made it possible for the field of phytogeography to emerge.

THE HISTORY OF PLANTS

The role of plants and the idea of natural regions were treated quite differently in the work of Karl Willdenow, the Berlin systematist and mentor of Alexander von Humboldt. Willdenow (1765–1812) was one of the first naturalists to think of plant geography as a separate field of inquiry within botany. With his strictly pragmatic approach he narrowed the scope of distribution studies, thereby divorcing the subject from the wide range of arts and sciences invoked by Johann Forster. (Later, Humboldt was to expand the scope again.) Willdenow also released botanical geographers from any obligation to search for great unifying causes like heat. Such narrowness would perhaps have been restrictive had he not reintroduced the old question of the origins of the plant kingdom and sought to answer it with the concepts of the period. Like Forster, Willdenow found the idea of natural organic regions to be compelling; unlike Forster, he was concerned only to reveal the processes that led to the formation of these provinces, the history that governed the arrangement of the world as he knew it.

Through his friendship with Humboldt, Willdenow came to pay special attention to foreign floras, particularly when his pupil forwarded collections made during an expedition through Spain and asked for help in cataloguing the botanical results of his great voyage through Latin America. As head of the Berlin botanical garden from 1801 to his death in 1812, Willdenow worked hard to make it the most comprehensive collection of living and dried specimens in Europe.[4] In 1792 he published a justly famous primer for botanical students, the *Grundriss der Krauterkunde*, designed to replace the aging *Philosophica Botanica* (1751) of Linnaeus and later translated into English as the *Principles of Botany* (1805). In this basic textbook he outlined the elements of botanical geography, which he called the "History of Plants," using the word "history" in its older sense of description or "story."

Since plants fell into distinct geographical groups, Willdenow recognised the need to investigate the regularities of distribution. The underlying factors were seen to be at once physical (such as temperature, soil composition and structure, and the amount of light received), geological (that is, historical), geographical (height above sea level, nearness to ocean borders or lakes, and so on), and lastly social, including such phenomena, for instance, as the longevity of seeds and the tendency of bamboo to be gregarious. "By the history of plants," he wrote, "we mean a comprehensive view of the influences of climate upon vegetation, of the changes which plants most probably have suffered during the various revolutions this earth has undergone, of their dissemination over the globe, of their migrations, and lastly of the manner in which nature has provided for their preservation."[5]

Using these guidelines, Willdenow established five principal floras in Europe: the Northern, Helvetic, Austrian, Pyrenean, and Apenninian. In addition, he observed that the number of plant species—and hence the internal diversity of each assemblage—depended on the general temperature. Over the whole globe "the warmer the climate, the greater must be the number of wild growing plants." He demonstrated this point by listing collections from the Berlin garden: the total number of species in floras from different latitudes went from only two species in Antarctica (South Georgia Island), to 534 in boreal zones (Lapland), and to 4,000 in the tropics (Coromandel).[6] Like Forster and Buffon before him, and like Humboldt after his time, Willdenow was fascinated by the surface temperature of the earth and its influence on vegetation, but he was reluctant to pursue the topic in print. In his opinion botanical geography required facts more than it needed theories. He felt that the best thing to do, under the circumstances, was to concentrate on providing accurate figures and data for this as yet undeveloped botanical study.

Although his treatment of phytogeography was undoubtedly sparse and limited in scope, Willdenow did realise both the necessity for this study and the potential it held for a fuller understanding of the laws of creation. In the same section as the "History of Plants" he put forward an ingenious scheme for the origin of the plant kingdom that also explained its present topographical arrangement. At an early stage in the earth's formation, he suggested, it consisted of many mountains surrounded by a shallow but global sea. Plants were created on the peaks and gradually spread from these foci as the seas subsided. Each botanical province of the present day was the result of dissemination from a single mountain or mountain range—a suggestion that neatly explained the self-contained identity of such populations. To account for some of the oddities of distribution, such as the occurrence of *Ericaceae* in northern Europe and the southern tip of Africa and nowhere in between, he postulated a subsequent era of gales, volcanoes, and dramatic fracturing of continents, which carried some species way beyond their natural limits to the "furthest corners of the globe." This period of intense activity also served to explain the existence of fossilised plants, particularly "exotic productions," by which he meant tree ferns and the like, in the geological formations of Europe.

Willdenow's hypothesis therefore incorporated features of the universal flood and an epoch of chaos, which were evidently still essential to a Protestant interpretation of the past. His ideas are notable for their synthesis of elements drawn from the recent and not so recent history of botanical investigation. To him, botanical geography held the key to the creation and subsequent development of the plant kingdom. Yet he framed his explanation in conventional terms of a flood, isolated foci of species, and their eventual migration over the face of the earth. The only difference between his views and those of Linnaeus—important though it is—is that Willdenow had consciously dropped the Garden of Eden and was thinking not of one point of origin for all living beings, but of many. His scheme contained as many aboriginal mountains as there were modern botanical provinces. Should a new province become known to science, all that had to be done was to postulate another peak: like many cosmogonists, Willdenow felt no need to specify real locations. He could add or subtract a mountain as and when the occasion demanded. Ad hoc as it may seem, the theory was nonetheless a genuine attempt to relate his new field of the "history of plants" to deeper, more philosophically relevant questions about the origins of the natural world.

Willdenow's efforts did not go unrewarded. During the last decade of the eighteenth century and the first years of the nineteenth, there was a deeply metaphysical structure to the study of plant geography, and the first self-styled "botanical geographers" deliberately pursued this topic as a tangible and readily accessible path toward understanding nature. Willdenow's successors believed that the topographical patterns found in the natural world held important clues for the knowledge of life itself. These men, among them Alexander von Humboldt, Augustin de Candolle, and Robert Brown, believed it essential to describe and delimit particular living associations as a first step to ascertaining the environmental factors specific to each area; the more they knew about the organic and physical characters of a region, the more they anticipated the disclosure of rules or generalisations about distribution patterns. In turn, these generalisations were expected to reveal something about the mystery of life and the causes of nature's variety.

There were good reasons for this enthusiastic commitment to the "history" of plants. In part a reflection of more general trends in the sciences, which were then shifting to focus more clearly on forces and origins, it was equally an outgrowth of post-Linnean natural history. Once floras and faunas had been recognised as discrete associations, it was perfectly unexceptional for scientists to investigate the reasons why. Once biological provinces and the idea of groups had been brought into the conceptual equipment of naturalists, it was only reasonable to deploy these units in the search for more fundamental regularities. Inquiring minds seized upon these notions, only recently made clear to science, and began to utilise them.

Furthermore, at a deeper cognitive level, there was a strong bond between the study of distribution on the one hand and a search for the causes underlying nature's diversity on the other. Although perhaps it had always existed in one form or another, that bond was first made explicit in the Linnean species concept. After Linnaeus had published his views, the notion of "species" was twofold or bipartite, thereby providing later naturalists with two possible courses of action in the pursuit of their intellectual goals. The concept of species consisted of, first, an idea of the differences in *kind* or structure and, second, an idea of the geographical position or *place* of that body, mediated by the concept of "suitability"—that each organism was ideally suited for its specific place in nature. Therefore to understand fully the meaning of living beings it was as legitimate to study one aspect of species as it was to study the other.

Naturalists could choose to examine the structural differences between beings of all kinds and attempt to isolate the regularities of form which must have been the Creator's (or nature's) "building-blocks" or which were quintessential elements of nature itself. This approach can be seen in the work of *Naturphilosophen* comparative anatomists, particularly in the writings of Lorenz Oken, Henri de Blainville, and Goethe, who researched into the problem of ideal types and the structure of metaphysical archetypes.[7] Another aspect of the same "anatomical" or "structural" approach to nature's phenomena was championed by those French physiologists and experimental biologists who attempted to document the influence of local environments on organic form. Attention was directed to the changes in structure brought about by altered surroundings. Taking their cue from Buffon, who had suggested that American animals degenerated according to the environment, botanists such as Adanson and Duchesne considered the mutability of plant forms under different external pressures. Lamarck too, as is well known, considered the interrelationship between inorganic factors and living beings, while Étienne Geoffroy Saint-Hilaire studied experimental teratology in an attempt to substantiate the same relationship during embryonic development. Together, the ideas of these men reflected a general concern with the underlying reasons for difference in "forms" and "kind."

Naturalists could also search for the meaning of nature along alternative routes, notably that of the geographical arrangement of living beings. Here the idea of "kind" was replaced by that of "place." Geographical location was an equally important part of the post-Linnean definition of species; it should therefore be able to provide an account of nature's essential properties just as readily as comparative anatomy or organic mutability. To Alexander von Humboldt this notion was compelling. Apart from any intrinsic interest that the subject of geographical distribution might hold for science, the main reason for pursuing this study was, he felt, because it could lead to the disclosure of fundamental natural laws. In his opinion, the study of distribution was an inquiry of the highest philosophical order. It allowed naturalists to reflect on the original creation of living beings, their outward appearance, and their situation in life; it explained why nature existed in the way that it did.

HUMBOLDT AND BOTANICAL GEOGRAPHY

Alexander von Humboldt (1769–1859) is remembered, among other things, for turning the study of plant distribution into a full-fledged scientific endeavour; he brought to his interest in plants a zest for physical information and a love of the grand and sweeping overview. It was Humboldt who outlined the scope of the inquiry, and it was he who gave naturalists their first concrete rules for procedure.

During his youth—spent acquiring the best general education that Europe could provide—Humboldt had been greatly impressed and influenced by his friendship with Johann Forster's son Georg. Georg Forster had taken Humboldt with him on an excursion in 1790 through the Low Countries and on into England, where they examined the social history, politics, and language of the countries visited, after which they boasted of having seen every notable display of art, antiquities, and curiosities that public and private collections could offer.[8] The younger Forster encouraged in Humboldt a deep appreciation of the humanist approach to nature and mankind, a philosophy that both men had encountered during their times at Göttingen and Berlin. Forster—a profoundly restless and polymathic scholar—was at that time at the height of his short-lived fame, having published significant tracts on German literature and political theory. He was also noted for his attempts to bring the French Revolution to Germany.[9] Forster strongly opposed any contemporary specialisation in the sciences or arts; in this he was influenced first by Buffon and other Enlightenment thinkers who presented a total view of nature and then more markedly by the educational programme he had followed as a student. Humboldt was always to remember the way in which his companion endeavoured to foster these feelings in his own mind. As it turned out, he did his work well: Humboldt went on to surpass Forster in every field, writing one of the most synthetic studies of the age, the *Cosmos*, a "sketch of a physical description of the universe," in 1848.

Under Forster's tutelage, the younger man developed a vision of the interrelated nature of the universe in which plant geography played a small but not insignificant role. The study of plant assemblies was, for Humboldt, a recapitulation in miniature of the study of one magnificent whole; the method was the method of all intellectual inquiry; the subject matter enriched and illuminated the world of men and aesthetics.

In a few seminal tracts published between 1807 and 1817, in which he dealt with the botanical results of his important travels through Latin America (1799–1804), Humboldt drew together all that had previously been written on botanical geography and added his own observations in order to set this new investigation on its feet. Never a modest man, he also made quite sure that his readers knew that this was what he was doing. He noted that he—a great thinker of the new age—was single-handedly creating an important science. "I consider it a happy circumstance," he wrote at the end of his life, "that at the time during which I devoted my attention almost exclusively to botanical pursuits, I was led by the aspect of the grand and strongly characterised features of tropical scenery, to direct my investigations towards these subjects." Humboldt alone could provide the "comprehensive glance" that would reveal the "natural physiognomy" of the earth.[10]

Humboldt's most famous work in this respect was his elegant *Essai sur la géographie des plantes* (1807). According to his memoirs it was first sketched out in 1802 at the foot of Chimborazo, a mountain then thought to be the highest in the world and which Humboldt proceeded to climb. From the plains of Tapia he calculated its height as 18,096 feet—a record for mountaineering endeavour until Joseph Hooker, another adventurous botanist, reached 19,000 feet in the Sikkim Himalaya in 1848. The *Essai* inspired in this dramatic fashion was subsequently prefixed to an elaborate survey of the physical, meteorological, and geological characteristics of equatorial mountains, measured at different heights. These factors were correlated with changes in the vegetation according to elevation, giving rise to Humboldt's notable discussion of the "physionomie" of each separate assemblage of plants. Here it was not so much the identity of the species occurring in different zones of elevation that mattered as the overall appearance of vegetation at succeeding heights. Differences in the environment were reflected by the sort of organisms that could be supported there, and so, in passing from the foothills of an equatorial mountain to its summit, Humboldt noted a succession of markedly different communities of plants. These were analogous in form to the gatherings found on other mountains and analogous to groups of plants found in succeeding belts of latitude. In other words, Humboldt claimed that he could see on the slopes of Chimborazo a complete series of floral belts typical of the tropics, of the temperate regions, and of boreal or arctic latitudes. At the base, tree ferns and palms were to be found. Higher up were broad-leaved forests, superseded by deciduous woodlands often typified by birch. In turn, these were replaced by coniferous forests, which gave way at greater heights to grassland and dwarf shrubs. This last diminished according to the severity of local conditions until at the very crest of the mountain Humboldt expected to find bare rock and scree, accompanied by permanent snow if the peak were high

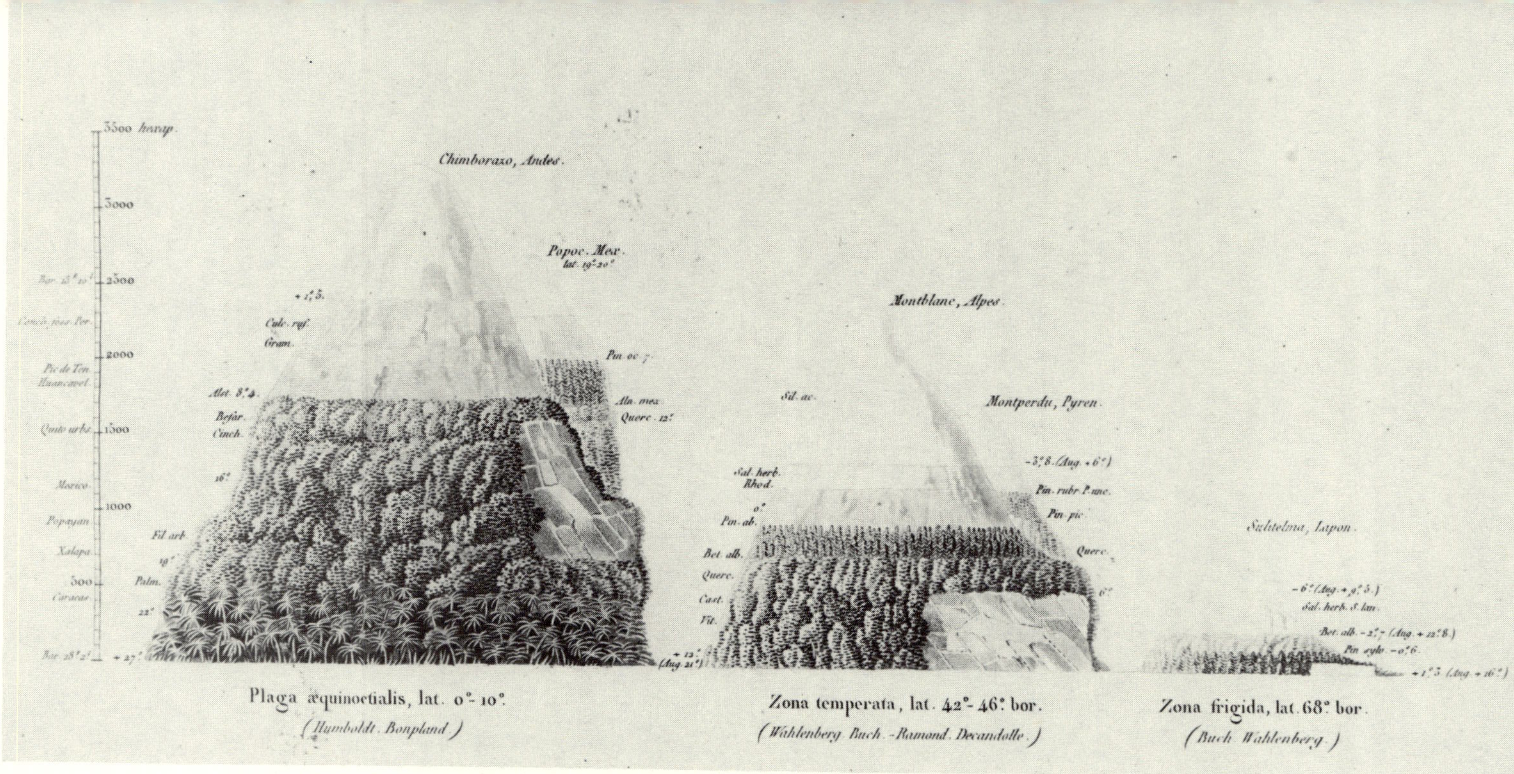

Illustration 3 Humboldt's diagram of the geography of plants. Mountains in tropical, temperate, and boreal latitudes have their vegetation arranged in analogous belts. Altitude was the critical factor, here measured in six-foot units. From A. von Humboldt, *De distributione geographica plantarum* (Paris: Libraria Graeco-Latino-Germanica, 1817), frontispiece.

enough. A mountain was therefore a hemisphere in miniature, where the floral zones of the earth were found repeated on a vertical scale, so that "climates ranged, as it were, in strata one above the other."

It was only in the most general of terms that Humboldt could draw this kind of analogy between distribution according to latitude and to elevation, and his use of the word "physionomie" to characterise each association was correspondingly generalised. "Physionomie" denoted a "type" of vegetation, such as grassland, palm groves, or temperate forest, which was, for the sake of convenience, often named after one of its more conspicuous species. In the *Essai* of 1807 he listed fifteen types of vegetation illustrating different overall "form," and he insisted that it was not the presence of the named family—the conspicuous species that gave the collection its name—which was significant, but rather the essential appearance of the whole "assemblage." Was it lush and full of different botanical orders, or was it made up of one predominant life-form, such as birch or pine trees? Was there diversity or "monotonie" in the spectacle which presented itself to the human eye? Humboldt's interests—always open to aesthetics—lay in the comparison of the "harmonie" and the "beauté absolue" of forms, which correspondingly revealed the underlying character of the various regions of the globe.[11]

The most important insight which Humboldt derived from this study, and which comprised the gist of his message in the *Essai*, was that groups of plants should be considered as coherent entities, some of which mimic each other. On a mountain, every floral belt was a collection of species that reflected some other collection typical of certain latitudes. On the flat, the floras of the northern hemisphere imitated those of the south, constrained only by the relative lack of land and the peculiarities of climate found in the austral regions. Everything was related to everything else, yet continued to retain a certain quality that gave each community its own particular identity.

This sweeping view of the geographical arrangement of plants was intimately related to Humboldt's concern with the physical characteristics of the globe. He saw living "assemblages" as striking guides to the climate and physical conditions in which each naturally occurred. The analogy, or repetition, of similar "types" over topographical space demonstrated in a trenchant manner the regularities that he believed underlay the physical nature of the earth, regularities that could be mapped and charted to produce global patterns that would reveal something about the ultimate meaning of life and the cosmos. The arrangement of plants was clearly governed by a series of factors, ranging from general location according to latitude—and hence according to the surface temperature of the earth—to the local effects of altitude, nearness to the sea, or the chemical composition of the supporting rocks. All these factors, implicit in Johann Forster's work and first listed by Willdenow, were brought into the open by Humboldt. They could be measured or otherwise determined in a numerically precise fashion, tabulated, and arranged in an attempt to pinpoint the crucial combination giving rise to the group in question. Accordingly, Humboldt invented the isobar—a line on a map joining points of the same atmospheric pressure—and the isotherm, which did the same thing for minimum or maximum average temperatures during the year or in one particular season, such as the growing season in the arctic circle. These lines formed physical boundaries beyond which certain types of plants or their characteristic assemblages could not be found.

The aim of Humboldt's botanical geography was to compare and contrast plant assemblies according to their physical characteristics and locality. "Botanical geography," he wrote at the head of the *Essai*, "is a science which considers vegetation under local conditions in different climates." The best guide to the breadth of Humboldt's interests in this respect is the wide range of measurements that he personally considered essential for a truly scientific account of the flora of Latin America. He presented a tableau encompassing "la végétation; les animaux; les rapports géologiques; la culture; la témperature de l'air; les limites des neiges perpétuelles; la constitution chimique de l'atmosphère; sa tension électrique; sa pression barométrique; le décroissement de la gravitation; l'intensité de la couleur azurée du ciel; l'affroiblissement de la lumière pendent son passage par les couches de l'air; les réfractions horizontales, et le degré de l'eau bouillante à différentes hauteurs."[12]

Humboldt felt that the plant life found in any region was very much a feature of the particular circumstances that region had to offer. Like his colleagues Willdenow and the two Forsters, he believed that physical surroundings were the prime cause behind the distribution of species into easily demarcated groups, although unlike Willdenow he concentrated primarily on existing patterns and the relationships between today's associations rather than concerning himself with their first appearance on the globe. His views were more akin to those of Johann and Georg Forster, who held that there were intimate links between the inorganic features of an area—including its geographical size and location, topography, soil and mineral wealth, and the nature of the climate—and the organisms it contained. The plants, animals, and men inhabiting tropical islets portrayed in their structure the particular qualities of their surroundings, just as clearly as the inhabitants of bleak Tierra del Fuego portrayed the harshness of their environment. For the Forsters, the nature of the organic world and human society changed from being "beautiful, rich, enchanting" in the tropics to "deformity, poverty and disgustfulness" in the polar landmasses of the world. Humboldt echoed this environmentalist opinion, relating even the artistic prowess of nations to their physical surroundings: "How powerfully have the skies of Greece acted on its inhabitants! . . . The poetry of the Greeks, and the ruder songs of the primitive northern nations owe great part of their peculiar character to the aspect of the plants and animals seen by the bard, to the mountains and valleys which surrounded him, and to the air which he breathed."[13] All human life was thus influenced by the environment, and Humboldt's plant assemblages acted as intermediaries between the physical and the "moral" worlds. The results of botanical geography, he stated, "comprise in one view the climate, and its influence on organised beings, the aspect of the country, varied according to the nature of the soil and its vegetable covering, the direction of the mountains, and the rivers which separate the races of men as well as the tribes of vegetables, and finally, those modifications, which the state of nations, placed in different latitudes, and in circumstances more or less favourable to the display of their faculties, undergoes."[14] Botanical geography was an important subject indeed.

THE NATIONS OF PLANTS

In a way this environmentalist view, common to Humboldt and the Forsters and their interest in the physical properties underlying the different biological provinces of the world, was an outgrowth of intellectual movements in eighteenth-century France. There, Buffon and others had emphasised, first, the regionality of the globe and, second, the significance of the environment. These observations were rapidly assimilated into contemporary thought, stimulating and intriguing research biologists. At the newly reconstituted Jardin des Plantes, now known as the Musée d'Histoire Naturelle, Étienne Geoffroy Saint-Hilaire was busy investigating the effects of environmental changes on embryonic animals while Lamarck worked out the details of his system of organic transmutation according to the influence of individual surroundings. Yet their interests concentrated on questions of definition, on the nature of species, and how differences in form were generated and maintained in the real world in preference to the global patterns that preoccupied Humboldt and his German colleagues. It was only in a German context that all-embracing statements about natural regions were being made, and it was only men who thought of themselves as "geographers" who studied plant distribution on a worldwide scale. Insofar as the investigation of macrodistribution was concerned, a centre of activity had appeared in the university town of Göttingen.

In Göttingen, scholars were trained in what might be called the liberal arts. The university had been founded in 1734 by the then Elector of Hanover (later to become George II of England) and was designed to foster the pragmatic tendencies of the age. It soon excelled in science and in the humanities as concerned with the problems of government, political science, and the arts of war. It boasted the first professor of geography in Europe, Tobias Meyer, who introduced subjects that would now go under the name of economic geography: the study of national finances, economic strategies, and the topographical limits on trade. These topics reflected Meyer's belief in the existence of distinct natural regions that had nothing to do with political boundaries. The university also formed the first centre of anthropology under J. F. Blumenbach, who specialised in the comparative measurement of skulls and developed a splendid collection of same for the study of physical anthropology. Blumenbach insisted on the subdivision of the world's inhabitants into certain aboriginal races, which had disseminated from their primary country. The conditions in which they then found themselves stimulated considerable modifications in form. The races of mankind, he stressed, were great natural groups, impossible to distinguish by mere political boundaries alone; like Meyer, he thought that political states and dominions were man-made units superimposed on some more primitive framework.

All this, plus courses in the arts of government, went under the general name of *Staatenkunde*—the craft of statesmanship. The professors at Göttingen elaborated on and refined the tradition passed on to them by the very earliest commentators of ancient Greece, who examined the political organisation, people, customs, produce, and industries of the countries of the known world in order to produce a science of government that was eventually to develop into full-blown political economy. Students were therefore educated in "statistics," a subject which denoted the collection of facts that might be of interest to a statesman or learned individual and which meant—quite evidently—the comparative description of states.[15]

The intellectual exercise of describing states in such a way can be traced back as far as Aristotle, whose *Politeiai* contained monographs detailing no fewer than 158 diminutive Greek nations. And, naturally enough, the awakening of political life in Renaissance Italy gave rise to similar compilations relating the pertinent facts of every city-state. But the subject only reached its fullest development with the activities of German authors of the eighteenth century.

At Göttingen the polyhistor Gottfried Achenwall established the content of *Staatenkunde*, defining it as the description of the constitution of states, a description that was to include all the information available. Achenwall consequently laid out the foundations for a subject of enormous breadth; his own work on the Spanish nation, published in 1748, was of considerable scope and encompassed the history, geography, climate, natural produce and "industry" of the country, the population and its distribution, the national character, political constitution, the present state of the sciences, arts, and financial affairs, and a short sketch of Spanish military and naval forces.[16]

So, when Johann and Georg Forster came into contact with the intellectual milieu at Göttingen, there was already a well-established school of thought having as its focus the relations between natural regions. Both naturalists absorbed this teaching with enthusiasm: for them, the floras, faunas, and physical surroundings of each area were all essential parts of a generalised description of a greater whole.

By the end of the eighteenth century, German intellectuals had further elaborated on this study of "nations." Men such as Goethe and William von Humboldt (Alexander's elder brother, a political thinker and leading statesman) were convinced that every nation or clearly defined group of individuals possessed a "characteristic," an essential quality that was special only to that particular assemblage. The Humboldt brothers, inspired by Goethe and Schiller, with whom they once associated on the closest of terms, believed that the characteristics of nations—or of any group—could be disclosed only through the almost intuitive divination of basic forces; there was a need for a special talent or "geist" on the part of the investigator to search beneath the surface attributes of nature for great—perhaps universal—themes.[17]

The search for essential qualities lying hidden below nature's surface phenomena preoccupied men of the Göttingen "set," who formed satellite groups in Jena and Berlin during the last years of the century. They delved into their own personalities in order to develop a refined sensitivity to the forces that lay beneath the surface of life and applied this instrument to all the varied topics that came within their range: Goethe discerned regularities in subjects as diverse as the comparative morphology of plants and the universal predicament of man; William von Humboldt sought to define the essential characteristic of the French nation in its revolutionary fervour; and Schiller reflected on the passions that ran beneath the flow of history in his great prose dramas, typifying the "Sturm und Drang" literary movement of the time. For his own part, Alexander von Humboldt saw this kind of investigation as one of the most ennobling and important contributions to human thought. He recognised in Goethe's work the quality of stripping nature to the core to reveal basic forces and unifying models for seemingly complex phenomena, like the anatomy of flowers or the morphology of the vertebrate skull. Humboldt too wished to bare the "archetypes" of nature and it was this urge that led him to investigate the outer world in the way that he did. Indeed, in a dispassionate assessment of his brother's skills, William von Humboldt claimed that Alexander was especially gifted in this sixth sense, the "geist" that he and his colleagues strove to attain, and was destined to produce work of the highest philosophical order.[18] Alexander did not disappoint him, and went on to write his own tribute to the underlying elements of the natural world in the *Cosmos* (1848), a book that was the peak of his personal development but, by the time of its publication, a somewhat outmoded style of thought.

Humboldt's work on phytogeography should be understood as only a small part of a far greater project to map out the characters of the world and those of its inhabitants. Nevertheless, the study of the minor topic was conducted along the same lines as the larger endeavour. He attempted to reveal the basic types of vegetation to be found on the earth and struggled to present their characteristic features in a way that simultaneously defined and typified each assembly. Furthermore, he was interested in searching out the unifying themes—the analogies—which interlaced these separate groups. He treated plant associations almost as if they were human "nations," comparing and contrasting their properties and relations within the group and outside it, selecting the one or two essential "characters" that not only made each group unique, but that linked them all together at a more basic, global level. Humboldt was the statesman—the "statist"—of the vegetable world.

HABITATIONS AND STATIONS

The detail Humboldt advocated and illustrated in his writings was further advanced by Augustin de Candolle, the first of the distinguished line of Swiss naturalists to bear that name. Candolle too was absorbed by the problems of plant distribution and gave his opinions to the French reading public in a justly appreciated little essay, the *Essai élémentaire de géographie botanique* of 1820. First published in the *Dictionnaire des Sciences Naturelles*, edited by F. C. Levrault, it was issued separately in the same year from Paris and Strasbourg. From this source the young Charles Darwin learnt his botanical geography, as did his friend and botanical guide Joseph Hooker, and, no doubt, the younger Candolle, Alphonse, who was to write on the same subject himself. All three quoted it extensively before going on to outstrip their teacher in the same field.

Augustin de Candolle (1778–1841) was an avid botanist from his earliest years and had been encouraged to take an interest in physiology and experimental biology by two local pastors, Jean Pierre Vaucher and Horace Bénédict de Saussure, themselves noted botanists and the latter one of the more illustrious Genevan scholars of the eighteenth century. Candolle was sent to study medicine in postrevolutionary Paris and remained there throughout the Napoleonic period, devoting his researches to botany under the guidance of Lamarck (then at the height of his fame), eventually revising and expanding Lamarck's useful catalogue of the *Flore Française* in 1805. Candolle focussed on the structural organisation of plants, their physiological requirements, and on a new branch of study then called phytochemistry—much encouraged by Lamarck—and medical botany. He played an active role in French scientific society, joining, in the aftermath of the Terror, the Société Philomatique, where he met Cuvier and collaborated with Biot on the thermal conductivity of gasses. More significant, however, was his membership of the exclusive Société d'Arcueil, founded by Laplace and Bertholet at the end of the 1790s.[19] Candolle was one of the original nine members, a glittering group which included Humboldt, then residing in Paris while writing up the many volumes of his *Voyages aux régions équinoxiales*. Humboldt's friendships among the Société d'Arcueil had drawn him to Paris after the South American journey, and kept him there in subsequently difficult years (when he was accused of being a Prussian spy, among other things). It seems likely that the same personal ties prolonged Candolle's visit to France: the Society, never larger than fifteen, encouraged the closest contact and intellectual exchange between its members, who resided for weeks at a time in the country homes of Laplace and Bertholet. As a group, they concentrated mainly on problems of experimental science relating to physics and chemistry, particularly on the properties of the atmosphere and light. In this respect, Candolle applied artificial illumination to plants in an investigation of photoperiodicity and heliotropism. These general topics were especially suited to Humboldt with his interest in the chemical composition and physical attributes of the atmosphere. The two became close friends.

As a result of his Arcueil connections, Candolle was given the task of travelling through the many regions of France still unexplored by botanists, with the utilitarian motive of recording connections between the botany, agriculture, and local conditions of each area. Although the appointment was held at Montpellier, he managed to maintain contact with Paris until, in 1816, he decided to return to Geneva, which had received independence from France and become a canton of Switzerland. There, he took up the chair of Natural History at the university and devoted himself to writing, teaching, and improving the facilities for botanical research, until his death in 1841.

The elder Candolle's ideas about plant distribution were an interesting mixture of French concepts about the dominating effects of the environment and a quasi-Humboldtian vision of global patterns and worldwide relationships between floras. All his Parisian researches indicated that plants were dependent on their environment, to the extent of their being influenced by the numbers of individuals or species contained within a particular region. Candolle was the first naturalist to make public the notion of a "struggle for existence," pointing out how individuals compete for space and light, or for other necessary resources, so that it might be said that "all nature was at war" with itself.[20] Over and above these important "sociological" and "economic" considerations, Candolle also stressed that the distribution of plants was governed by the physical conditions that each and every area did—or did not—present. He dubbed this the study of physico-geographical factors and elaborated on the topic in two notable essays, the "Mémoire sur la géographie des plantes de France, considerée dans ses rapports avec la hauteur absolue" (1817),[21] which took its impetus from

Humboldt's work on the effects of elevation on vegetation, and his "Projet d'un flore physico-géographique de la vallée du Leman" (1821).[22] Every plant species had a particular combination of requirements and could therefore be expected to appear only in the places that presented them in the appropriate manner.

Candolle used this assertion as the baseline for his more famous work, the *Essai élémentaire de géographie botanique*. Here, he divided "géographie botanique" into three sorts of investigation which should, the author stressed, be pursued as an integrated whole. The first section, concerned with the physical factors of temperature, light, water, inclination of the sun, and the atmosphere, was intimately connected with the second line of investigation, that of describing "Stations" or, as modern biologists would call them, habitats, such as moorland, marsh, or coastline. The final section was given over to a description of what Candolle termed "Habitations," or the great division of the vegetable kingdom into botanical provinces. This last not only was the net result of the first two categories, but also encompassed one of the most abstruse and fundamental inquiries that Candolle thought scientists could pursue. The study of habitations involved the reasons living beings were confined or circumscribed to areas that appeared to be their native countries. "Tous ou presque tous les végétaux," he wrote, "livrés à aux-mêmes, tendent à occuper sur le globe un espace déterminé; c'est la détermination des lois d'après lesquelles se fait cette circonscription végétale, qui constitue l'étude des habitations."[23] The whole subject was fraught with difficulties, and depended on the most obscure topic in natural philosophy at that time—the origin of species. Nonetheless, he continued, should botanists uncover the mysterious and problematic laws of phytogeography they would prove to be of the highest interest to the natural sciences as a whole. Candolle clearly saw the glory that would fall on botany if it came up with an answer for the crucial question of the origin of species.

Candolle also introduced a subtle distinction between the factors that determined distribution on the small scale and gave rise to "stations" and those which gave rise to global patterns or "habitations." A station, he wrote, is the place where a plant occurs naturally, according to its physical nature. A habitation is its overall geographic position. "Ainsi," he explained, "l'habitation du Riz est dans l'Inde, sa station dans les marais."[24] The one was strictly physical, physiological, and environmental, whereas the other was of a more generalised philosophical nature. This differentiation was important for Candolle because it separated topics that were, in his opinion, based on fundamentally different premises. The study of stations was concerned with secondary causes and was, at root, fully explicable. The study of habitations, on the other hand, dealt with incomprehensible primary causes, such as the origin of species and the primeval arrangement of plants over the earth; of necessity, it was a difficult and perhaps insoluble field of investigation. By dividing the subject in this way, Candolle ensured that naturalists were made aware of the aims and limitations of their work and were sensitive to the different ways in which plant distribution could be considered. Moreover, by dividing it into two halves and giving each a name, he made both sides appear more important than before. In particular, he enhanced the significance of the study of botanical provinces and encouraged his readers to pursue this demanding branch of philosophical phytogeography.

Candolle's division of botanical geography was important in other ways too, for this dichotomy in the subject was something his contemporaries had not hitherto considered. Humboldt, for example, had been more interested in establishing formal links between the environment and vegetation *per se*, and although he smoothly adjusted his mental focus to study both small- and large-scale phenomena, he had examined neither the methodological foundations of his investigation nor the epistemological restrictions of his search for scientific causes. Candolle seems to have been the first naturalist to think of these, thereby becoming the first to differentiate between the study of small-scale and the study of large-scale distribution patterns. A worthy successor to Lamarck, he emphasised the effects of the environment on localised populations; a close associate of Humboldt, he also recognised the subdivision of the earth's surface into coherent botanical provinces. From the time of his *Essai élémentaire* of 1820, the two disciplines of ecology (the study of stations) and of biogeography (the study of habitations) began slowly but surely to emerge.

For all his innovation, though, Candolle was not prepared to elaborate on the investigation of habitations or "des régions botaniques" as he subsequently defined them. He had already noted in the *Essai* that this field depended on a knowledge of the origin of species and their subsequent history. In later passages he commented on the thesis presented by Willdenow (of which he disapproved) and on the scanty information being derived from palaeobotany by Sternberg and Alexandre Brongniart in Paris. The idea of a historical biogeography that reached back beyond the period of man's existence was difficult for him to visualise, and practically impossible to document with so few fossil materials on which to base acceptable hypotheses. In this way Candolle was very like his friend Humboldt. Humboldt, for instance, was fascinated by the relationship between human history and the dissemination of useful plants such as Indian corn or other cereals. When he cast his mind further into the past, however, he fell back on other people's explanations, believing with Cuvier that there had been successive epochs in the formation of the earth, each stocked with animals and plants appropriate for the different conditions. Many of the plants, Humboldt supposed, had survived the most recent catastrophe—the deluge—and recolonised the earth from refugia, which explained why there were species now living in the tropics that also occurred in fossilised form in European strata.[25]

Candolle's views on the past life of plants were much the same as Humboldt's. But he found the whole field so novel and unexplored that he preferred to assess only the methods by which present-day provinces could be itemised and described. To understand these phenomena thoroughly was the first requirement in a more generalised endeavour to discover the origins of species: the present, he felt, should be examined before theorists attacked the past.

In this respect, Candolle's ideas were broadly the same as those expressed by Humboldt and Johann Forster. Their interests lay not in how things came into being but rather in the relationships of organisms existing today. They wished to compare and contrast, to see how phenomena were mutually consistent, and to disclose patterns more fundamental than those on the surface. In essence, they took a geographer's view of nature, searching out the major topographical divisions of the globe, listing the physical conditions of each region, and linking to these the plant life supported there. Vegetation was believed to be an accurate indicator of the various subdivisions of the earth's surface: at the most basic level, it was an indication of annual temperatures and climates, of degrees of latitude and suchlike; at more sophisticated levels, it was an expression of variations in altitude or physical location in regard to coasts, or lakes, etc.; and at the most complex level of interpretation, vegetation was thought to reflect the qualities of each region through the diversity and number of its species, the monotony or variety of its overall appearance and the abundance of individual organisms. Plants were the living emblems of an earth divided into many zones and provinces.

Because they fell into natural assemblages, plants were as significant to the geographer as human populations, and botanical communities were easily likened to "nations" of men. The methods of early phytogeographical analysis were therefore the same as those of statesmen interested in comparative statistics. Furthermore, the *aims* of phytogeographers were identical to those of statists. Humboldt, Forster, and Candolle all wished to document the living topography of the world, to study the relations plants held to each other and to their geographical location, to discern and compare regularities in botanical distribution, and to draw out conclusions about the interrelations of the entire plant kingdom. They were concerned with the "political economy" of plants. Yet topics such as the history or the origin of species were by no means passed over. On the contrary, these scholars were anxious to report the ways in which phytogeography and the study of plant nations would lead to a fuller understanding of the ultimate laws of nature; indeed, they believed that plant distribution was a more promising field of inquiry for such discoveries than, say, comparative anatomy or experimental physiology. Moreover, since the work of Willdenow, botanists and philosophers were well aware of the importance of the history of plants and fully conscious of the need to construct detailed hypotheses for the development of the vegetable kingdom. The significance of the origin of species and the history of life on earth was not ignored, but positively emphasised by the work of Humboldt and Candolle. They were convinced that, although the study of botanical provinces had not yet disclosed the origin of species, they had succeeded in finding the most appropriate manner in which to search for it.

THREE

A Science of Patterns

Humboldt's greatest phytogeographical accomplishment had been to describe the way plants fell into natural associations or "assemblages" that could be typified by their general "physionomie." He grouped things that resembled each other into discrete units and searched for analogies between different populations. Like most natural philosophers, he thought that the essence of scientific practice was the connection and comparison of objective phenomena, followed (for Humboldt, at least) by the almost intuitive divination of natural laws and forces. He diligently applied this belief to phytogeographical researches, insisting that the geography of plants should be based on detailed analysis of floras and on the overall comparison of biological provinces. No general rules, he asserted, could emerge without comparison between one major botanical region and another.

Yet the idea that botanical—or zoological—regions could actually be compared led naturalists into something of a quandary. They were not well equipped to carry out such a comparison, having neither methodological conventions nor authoritative concepts to underpin their work in this particular direction. They had few practical guidelines and confined analysis to the simplest of procedures. Nathanial Winch (who wrote on the geographical distribution of plants in northern England)[1] did the obvious thing, for example, and merely listed all the species in area A and all those in area B, leaving it to the reader to extract whatever information he desired.

To some extent this cursory treatment of biogeographical information was useful, for a country with one hundred species was clearly different from another with two thousand. Moreover, areas with only a few species in common could be distinguished from those with many identical organisms. This straightforward mode of comparison was adopted by Augustin de Candolle in the *Regni vegetabilis systema naturale* (1818–1821) and by K. S. Kunth in considering the distribution of *Compositae* over the entire world. But such activity did not take botanists very far in their search for the meaning and rules of plant distribution; it merely told them how many species belonging to individual families were indigenous to this or that country. At this point Humboldt stepped in and declared that he had the answer. The thing to do was to use—he coined the term in 1815—the method of "botanical arithmetic."

More than any other phytogeographical technique, "botanical arithmetic" enabled naturalists to cope with innumerable unorganised facts and give a semblance of order to their jumbled thoughts about distribution. Under the aegis of Humboldt, Robert Brown, and Augustin de Candolle, it became an exceedingly popular methodological aid, soon percolating into zoological and palaeontological studies and eventually coming to dominate the practice of geographical distribution through the second quarter of the nineteenth century. The method of "botanical arithmetic" provided a focus, so to speak, for natural historians otherwise committed to a wide variety of interests and an increasing diversity of subject material. And although it was short-lived, the enthusiasm generated by this useful technique was instrumental in confirming the position of geographical distribution studies as a "real" scientific exercise. Humboldt's methodological advice came to form the cornerstone of an early-nineteenth-century science of distributional patterns.

BOTANICAL ARITHMETIC

Botanical arithmetic, as Humboldt explained it, was an elementary numerical technique that reduced absolute figures into statements of a proportional kind, which could then be arranged with others in a table. All that had to be done was to calculate the ratio of species presented by one group of plants (such as a family) to another (such as the total number of species present). The figures then served several purposes, in that the botanist could discover the predominant forms of life in each region—if any—and isolate the general relationships between the main classes or families. Alternatively, he was equipped to trace the relative incidence of certain plants, for example the grasses or laurel family, through several geographic zones.

Humboldt put these ideas forward in a variety of works written after his *Essai sur la géographie des plantes*. Personally, he thought his major contribution to this topic lay in two *Prolegomenae*, forewords to botanical tracts condensing the results of researches in Latin America, the *De distributione geographica plantarum* of 1817 and his *Nova Genera et Species Plantarum* (1815–1825). But further essays on the subject of botanical arithmetic can be found in a piece on geographical distribution for the *Dictionnaire des Sciences Naturelles*[2] and in the various editions and translations of the *Aspects of Nature* (1808). The *Prolegomena* to his long list of new species and genera discovered in tropical America addressed itself to the construction of tables of proportions. It was Humboldt's hymn to the arithmetical method. He compared, for example, the distribution of annual species (both monocotyledonous and dicotyledonous) in the tropics with those of temperate and northern zones and calculated the proportion of annuals in each region. Observing that annual forms were proportionally more widespread in temperate regions than in areas to the north or south, he concluded that these plants were evidently more suited to the conditions present in intermediate latitudes.[3] It is easy to see how such statistics could be multiplied and arranged in a table. In the same work Humboldt gave details of eleven natural families in France, Germany, and Lapland and worked out the proportions each group exhibited to the whole. In this way he showed how the family of grasses, the *Gramineae*, formed the same proportion of the whole in all three countries, even though the absolute numbers of species were widely disparate: 284 species of grass in France made up one-thirteenth of the flora; 143 species in Germany made up one-thirteenth; and 49 species in Lapland, one-tenth.[4] In the table illustrated here (figure 1), he worked out the ratio of individual families to the whole flora according to latitude.[5] The grasses in this table bear the relationship of 1:15, 1:12, and 1:10 in the tropical, temperate, and arctic zones, respectively, thereby constituting a significant part of the flora of each region. By contrast, typically equatorial plants, for example the euphorbias, decreased from a ratio of 1:35 to only one species in five hundred at the arctic circle.

Tables such as these marked for Humboldt the establishment of a numerical botanical geography. To become numerical was essential, in his opinion, if the study of distribution was to be based on indisputable facts. Precision and accuracy were Humboldt's watchwords, and if other sciences were amenable to arithmetical investigations then so should botanical geography. Moreover, since his primary interests al-

CLASSES ET ORDINES NATURALES.	RATIO CUJUSCUNQUE CLASSIS VEL ORDINIS ad totam copiam Phanerogamarum locis planis provenientium: In Zona æquinoxiali (Cal. med. 27°.)	In Zona temperata (Cal. 10°—14°.)	In Zona glaciali (Cal. 0°—1°.)	ADNOTATIONES.
Agamæ cellulosæ.	1 : 5	1 : 2	1 : 1	
Filices		1 : 60	1 : 25	Germ. $\frac{1}{48}$ Gallia $\frac{1}{73}$.
Monocotyledones.	1 : 6	1 : 4	1 : 3	
Cyperoideæ.	1 : 60	1 : 30	1 : 9	
Gramineæ.	1 : 15	1 : 12	1 : 10	
Junceæ.	1 : 400	1 : 90	1 : 25	Amer. bor. $\frac{1}{130}$ Gallia $\frac{1}{86}$.
Glumaceæ vel tres ord. præced.	1 : 11	1 : 8	1 : 4	
Labiatæ.	1 : 40	1 : 25	1 : 70	Amer. bor. $\frac{1}{40}$ Gallia $\frac{1}{24}$.
Ericinæ et Rhododendra. . . .	1 : 130	1 : 100	1 : 25	Amer. bor. $\frac{1}{36}$ Gallia $\frac{1}{125}$.
Compositæ	1 : 6	1 : 8	1 : 13	
Rubiaceæ.	1 : 29	1 : 60	1 : 80	Gallia $\frac{1}{73}$ Germ. $\frac{1}{70}$.
Umbelliferæ.	1 : 2000	1 : 30	1 : 60	Amer. bor. $\frac{1}{57}$ Gallia [illegible].
Cruciferæ.	1 : 3000	1 : 18	1 : 24	Amer. bor. $\frac{1}{62}$ Gallia [illegible].
Malvaceæ.	1 : 50	1 : 200	0	Am. bor. $\frac{1}{135}$ Gallia $\frac{1}{145}$ Germ. [illegible].
Leguminosæ.	1 : 12	1 : 18	1 : 35	
Euphorbiaceæ.	1 : 35	1 : 80	1 : 500	
Amentaceæ exclusis Casuarin. .		1 : 45	1 : 20	

Adjecimus differentias notabiliores quas offerunt Zonæ temperatæ boreales utriusque Continentis. In universum minus fides habenda numeris absolutis quam rationi qua crescunt vel decrescunt numeri a polo versus æquatorem aut ab æquatore versus polum.

Figure 1 Ratio of botanical classes and orders in the northern hemisphere. From A. von Humboldt, *Prolegomena, Nova Genera et Species Plantarum* (Paris: Lutetiae, 1815), p. xviii.

ways lay in the relations that the various parts of the whole held to each other and to the whole itself, this technique was ideal for constructing generalised statements about complex patterns. The entire study of phytogeography, he claimed, would be the better for using botanical arithmetic:

> Terrestrial physics have their numerical elements, as has the system of the Universe, or celestial physics, and by the united labours of botanical travellers we may expect to arrive gradually at a true knowledge of the laws which determine the geographical and climatic distribution of vegetable forms. . . . The numerical laws of the families of plants, the often striking agreement of the numbers expressing their ratios, where yet the species of which the families consist are for the most part different, conduct us into the mysterious obscurity which envelopes all that is connected with the fixing of organic types in the species of plants and animals, or with their original formation or creation.[6]

Yet Humboldt was not entirely cavalier about recognising his predecessors and contemporaries in this one field of botanical arithmetic. He conceded gracefully that both Robert Brown and Augustin de Candolle had approached the same theme from a different angle. In 1814 Brown had worked out the ratio that phanerogamous plants held to cryptogamous species over the various latitudes of the earth and, again, the ratio of monocotyledons to dicotyledons over the world[7] in order to determine what proportion of the total flora of any country a given collection of specimens might represent. The value of a herbarium, naturally enough, increases as the probability of its exhaustiveness rises, and Brown was taxed with the problem of assessing collections made "in a period not exceeding two months, and in a season somewhat unfavourable." Absolute numbers were of little account here. Brown saw the need to employ proportions as a kind of constant, from which his collections deviated more or less according to the quality of the botanising and which revealed the approximate number of species yet to be discovered. In another context, Brown also calculated the number of genera in certain families at various locations on the Australian continent. The region with the most genera was, in his opinion, the native home or country of that family. This was a particularly valuable device for working out the diverse relationships between the floras of Australasia, South Africa, and South America, and, on a more restricted scale, for suggesting the geographical source, or centre, for groups such as the *Proteaceae*.[8]

Candolle, on the other hand, moved onward from Brown's and Humboldt's usage to look at botanical arithmetic on a more precise level. He demonstrated in the *Essai élémentaire* of 1820 that a knowledge of the proportions between smaller taxonomic categories, between species and genera or species and families, could give equally useful distributional information. Such a computation was, of course, a mere extension of the practice of taking the ratio of families to larger sections of the vegetable world, but Candolle showed that the relationship between lesser classificatory groups was more useful than the proportions held by major groups. He worked out, for example, the criteria on which the overall diversity of floras could be established. The more species there were for every genus (say, 16 species to one genus on average), so the more constant and monotonous the plant covering. Correspondingly, the more genera relative to species, the more diverse the flora. To illustrate his meaning Candolle cited the example of France, which possessed an average of 7⅕ species per genus, whereas the British Isles had 2⅓, and the Canary archipelago, 1½. He wrote of these figures:

> Pour achever ce qui relatif à cette espèce d'arithmétique botanique, comme l'appelle M. de Humboldt, et pour montrer jusqu'a quel point elle peut peindre l'aspect général de la végétation des pays divers, je dirai encore qu'on a tiré quelque parti de la comparison du nombre proportionel des espèces et des genres d'un pays. Plus le nombre moyen des espèces de chaque genre ou de chaque famille est borné, plus l'aspect de la végétation présente de variété; plus, au contraire, ce nombre est grande, plus le coup d'oeil du pays présente de monotonie dans les formes.[9]

His "coup d'oeil du pays" showed that France had more species per genus than the Atlantic island group and consequently presented a "monotonie dans les formes." The Canaries, with only one and a half species for every genus (a one-to-one ratio is rare and very extraordinary when it occurs), evidently possessed a wealth of different sorts of plants, even though the total number may have been quite small.

This assertion might seem paradoxical, in that Candolle's monotonous areas appear to have *more* species (each of which, by definition, is a different form) than his diverse areas. But the crux of the matter is that there is a greater distinctness between genera than between species, and so it is their numbers which are the important element in estimating the overall type of vegetation. The frequency of higher orders of distinction, not the total number of closely related species, makes for richness. It was the notion of complexity that Candolle wished to convey—the presence or absence of extremes within each association. Actually, he might have achieved his aim more satisfactorily if he had calculated the same ratio the other way around; if he had demonstrated the average number of genera to 100 or 1,000 species, as his son Alphonse was to do. Ironically, a modern phytogeographer, P. Jaccard, has recently discovered the value of this alternative calculation.[10]

The index of diversity, if it may so be called, was independent of the absolute numbers of either species or genera. Candolle had created a method by which the vegetation of lands of different sizes could be compared without the presence of complicating factors, the most intrusive of these being an a priori assumption that areas of large topographical extent must possess a greater proportion of species than oceanic islands. Ever since Johann Forster's time it had been tacitly assumed that area and number of species must be directly linked. Recognising that, to a certain extent, this will often be the case, Candolle nevertheless showed that different relationships were also revealed by studying the ratio of species to genus. This ratio indicated fundamental distinctions in the kind of vegetation covering the globe: subsections of continental landmasses equal in size or otherwise comparable to islands usually possessed proportionally *fewer*, not more, species than islands.

THE NUMERICAL RELATIONS OF PLANTS

Candolle's method of calculating the number of species to genera turned out to be very popular. Christian Leopold von Buch used it to great effect in his essay on the organisms of the Canary archipelago, calculating the ratio of species to genera in both islands and the neighbouring continent and suggesting that the increased diversity of island floras was the result of occasional migration from nearby landmasses coupled with prolonged isolation.[11] The same ratio caught the imagination of the young Charles Darwin. In his species notebooks, written from 1836 to 1838 or so, Darwin tried to work out the differences between islands and adjacent continents, taking care to make calculations of his own for the Keeling and Galapagos groups, and noting also some figures from Humboldt and from Buch. He reflected that although "the flora of islands are poor," the number of genera was relatively high: "For instance, the temperate part of Teneriffe [*sic*]: the proportion of genera 1:1."[12] Equally, calculations such as these were so familiar to Joseph Hooker that when he wrote to Darwin about botanical distribution in many letters over the decade 1844 to 1854, he insouciantly assumed that Darwin would know all about the differences between "continental" and "insular" proportions. Hooker went on to base several important biogeographical arguments on these very differences.

One of the first protagonists of botanical arithmetic apart from the triumvirate who brought it into the public eye was Joakim Schouw of Copenhagen. His *Grundtraek til almindelig Plantegeographie* of 1822, with its German translation of 1823, was a valuable contribution toward understanding the distribution of plants in Germany. During his *wanderjahre* through that country and northern Italy Schouw had attempted not only to set in order the botanical results of his travels, but also to absorb the latest in phytogeographical teaching from Candolle and Humboldt, and returned to Copenhagen to take up the chair of botany with particular responsibility for introducing this new subject to eager pupils. His book contained a full discussion of all the criteria stipulated by Humboldt and Candolle, mostly in the form of charts and tables of factors such as temperature, pressure, and insolation. Numerical analyses were used to great effect, as, for example, in his estimation of the fraction of the whole flora held by certain plant orders. Yet Schouw's close links with both these eminent scholars of the Arcueil place him rather more in the position of enthusiastic disciple than as an equal and independent appreciator of botanical arithmetic; his work therefore has interest as one of the earliest attempts to apply—not invent—a new analytic technique.

After Schouw had set the example, other naturalists hastily followed suit. From the 1830s onward the "numerical method" became a regular and essential part of distributional surveys and was reinforced by the publication of elaborate tables. The technique proved so fashionable that even comparative strangers to the study of distribution valiantly attempted to use figures in an effective manner.

So, for example, Rudolf Wagner, the anthropologist, physiologist, and comparative anatomist of Erlangen, made a foray into numerical relationships, which seems oddly at variance with his immediate interests. In 1833 he compared the totals of specified groups of animals, such as mammals, birds, and insects, in successive authorities ranging from Linnaeus to J. C. Fabricius.[13] His primary aim was to demonstrate how rapidly the morphological sciences were advancing and, second, to illustrate the abundance of microscopic forms like infusoria in the various regions of Europe. Although his columns of figures were arranged to facilitate comparisons, and although he modelled his text on botanical surveys, he did not use ratios or proportions to illuminate his conclusions, since the expressed purpose was to discuss these in terms of absolute numbers.

But truly "statistical" estimates were also being published. In the same year of 1833 a young Edinburgh student, Hewett Cottrell Watson (1804–1881), made one of his earliest public statements in what was to become a long and distinguished botanical career. As a student of Robert Graham's he had early developed a taste for the field trips the latter had recently introduced at the university, paralleling similar developments at Glasgow under the direction of W. J. Hooker and at Cambridge under J. S. Henslow. Watson was particularly fond of longer excursions into the remoter and botanically unexplored parts of the Highlands. One such trip yielded the data on which his 1833 paper was based: "Observations made during the summer of 1832 on the temperature and vegetation of the Scottish Highland mountains in connection with their height above the sea."

Watson already possessed decided ideas about the proper way to study geographical distribution and had gone to the trouble and expense of having a small guide to these ideas privately printed in 1832, for distribution among his friends. The *Outlines of Geographical Distribution* paid special attention to vegetation at different heights, something that had rarely been investigated in Britain. The book also presented "a few illustrations of the arithmetical distribution of British species and families." These illustrations were considerably extended and formalised for his 1833 paper.

In that paper Watson listed and described the 306 species he found growing on Scottish mountains above an elevation of 1,000 feet. He separated these into groups according to height and subsequently arranged them into tables listed against the altitude at which individual families occurred. In other words, he displayed both the number of species from named "natural orders" (that is, families) at any one height and the spread of these orders over a range of elevation from 1,000 to 4,320 feet. In one table, for example, Watson calculated that the *Rosaceae*, with 20, 12, and 5 species at successive heights, formed a larger proportion of the whole population in each altitudinal zone than, say, the heathers (*Ericaceae*), which were usually considered to be at their maximum in the Highlands (figure 2).[14]

Watson was to make the study of the spatial arrangements of plants very much his own through his dictatorial dominance of (first) the ill-fated Botanical Exchange Club and (later) the newly formed Botanical Society of London. He bullied the Society into bringing out a standard checklist known as the *London Catalogue of Plants* (1844 and many subsequent editions), which he alone collated. His magnum opus, the *Cybele Britannica: or British plants and their geographical relations* (1847–1859), was undoubtedly one of the touchstones of topographical investigation from the time of its publication to 1859, when Darwin gave botanists evolutionary theory to think about instead. In the *Cybele* Watson devoted much space to "those arithmetical comparisons to which phytogeographers commonly seem to attach much importance." In a numerical account of the incidence—or absence—of species in precisely demarcated "Provinces" and "Vice-Counties," he managed to take the first step in a nationwide census of British plants. Accessory to this enterprise was his clear differentiation between mapping the individual points where each species had been sighted (in modern

TABLE of the Altitudinal Elevation of Highland Plants.

Natural Orders.	Numbers.			Proportions.		
	1000–2000 Ft.	2000–3000 Ft.	3000–4320 Ft.	1000–2000 Ft.	2000–3000 Ft.	3000–4320 Ft.
Ranunculaceæ	8	6	3	1/34	1/30	1/27
Cruciferæ .	9	7	5	1/30	1/26	1/16
Cistineæ .	1	...	...	1/273	...	...
Violariæ .	4	3	1	1/68	1/61	1/80
Droseraceæ .	3	1	...	1/90	1/183	...
Polygaleæ .	1	1	...	1/273	1/183	...
Caryophylleæ	9	12	9	1/30	1/15	1/9
Lineæ .	1	...	...	1/273	...	...
Hyperacineæ	1	...	...	1/273	...	...
Geraniaceæ .	2	1	...	1/136	1/183	...
Oxalideæ .	1	1	1	1/273	1/183	1/80
Leguminosæ .	13	7	1	1/21	1/26	1/80
Rosaceæ .	20	12	5	1/14	1/15	1/16
Onagrarieæ .	4	3	1	1/68	1/61	1/80
Halorageæ .	1	...	...	1/273	...	...
Portulaceæ .	1	1	...	1/273	1/183	...
Crassulaceæ .	1	1	1	1/273	1/183	1/80
Saxifrageæ .	7	9	8	1/39	1/20	1/10
Umbelliferæ .	5	...	...	1/55	...	...
Caprifoliaceæ	3	2	...	1/91	1/91	...
Rubiaceæ .	4	2	1	1/68	1/91	1/80
Valerianeæ .	1	...	...	1/273	...	...
Dipsaceæ .	1	1	...	1/273	1/183	...
Compositæ .	32	18	6	1/9	1/10	1/13
Lobeliaceæ .	1	...	...	1/273	...	...
Campanulaceæ	1	1	1	1/273	1/183	1/80

Natural Orders.	Numbers.			Proportions.		
	1000–2000 Ft.	2000–3000 Ft.	3000–4320 Ft.	1000–2000 Ft.	2000–3000 Ft.	3000–4320 Ft.
Vaccinieæ .	4	4	3	1/68	1/46	1/27
Ericeæ .	10	9	2	1/27	1/20	1/40
Gentianeæ .	2	...	...	1/136	...	...
Boragineæ .	4	1	1	1/68	1/183	1/80
Scrophularineæ	14	8	3	1/19	1/23	1/27
Labiatæ .	7	2	1	1/39	1/91	1/80
Lentibularieæ	1	1	...	1/273	1/183	...
Primulaceæ .	3	1	...	1/91	1/183	...
Plumbagineæ	1	1	1	1/273	1/183	1/80
Plantagineæ .	3	...	...	1/91	...	...
Polygoneæ .	7	3	3	1/39	1/61	1/27
Euphorbiaceæ	2	...	...	1/136	...	...
Urticeæ .	3	1	...	1/91	1/183	...
Amentaceæ .	14	10	2	1/19	1/18	1/40
Coniferæ .	2	2	...	1/136	1/91	...
Empetreæ .	1	1	1	1/273	1/183	1/80
Juncagineæ .	1	1	...	1/273	1/183	...
Orchideæ .	5	5	...	1/55	1/37	...
Melanthaceæ	1	1	...	1/273	1/183	...
Junceæ .	10	10	6	1/27	1/18	1/13
Cyperaceæ .	18	20	7	1/15	1/9	1/4
Gramineæ .	25	13	7	1/11	1/21	1/11
Total of Sp.	273	183	80			
Total of Ord.	48	38	25			

Figure 2 H. C. Watson's table of the altitudinal elevation of Highland plants. From H. C. Watson, "Observations made during the summer of 1832 on the temperature and vegetation of the Scottish Highland mountains in connection with their height above the sea," *Edinburgh New Philosophical Journal* 14 (1833): 323.

terminology, point maps) and fixing on an area within which given plants might be expected to appear. Watson devised "a test which is adapted to show the number of *spaces* (not the number of precise *places*) in which each one occurs."[15]

The *Cybele* contained many well-reasoned statistics of "geographical relations," including Watson's special love, that of altitudinal variations in plant groupings, but also many other interesting and innovative ratios. For example, he calculated the acreage over which any one species ranged, coming up with an average figure of 32.16 square miles for every species in eastern Britain. Additionally, he was keen to ascertain how many species congregated in different localities and worked out that any unit of ten square miles in the east of Britain would carry roughly one-third of the entire regional flora.[16]

In Germany, too, the "simplest reckoning" was duly appreciated. The Prussian botanist and plant physiologist Franz Meyen—a protégé of Humboldt's—wrote in 1836 a *Pflanzengeographie* in which a full thirty-page section was given over to the "Statistics of Plants." This work was later translated by the Ray Society and published in 1846 for the edification of British naturalists; both Darwin and Hooker owned well-thumbed copies and congratulated each other on the existence of a Society sophisticated enough to provide such translations.

For the benefit of those who did not know, Meyen explained the use of numerical data: "the method . . . is founded on the simplest reckoning, by which we find out the number of species of the smaller, as well as of the larger groups of plants, and compare the numbers obtained with each other, or with the whole number of all the species of a country." And lest the advantage of doing so still evaded the reader, he stressed that statistical enquiries into the "absolute and relative numbers of species" would lead to the revelation of fixed laws relating to botanical distribution. This, he pointed out, was a good thing.[17]

STATISTICS OF DISTRIBUTION

Of course, interest in arithmetic was not confined to botanical studies alone. There was comparable activity in zoological and palaeontological circles throughout the second quarter of the nineteenth century, modelled on the example of botanical arithmetic. Several zoological authors expressly maintained that they were applying the methods of phytogeography to their own branches of the living world. Some years earlier Humboldt had anticipated this transfer from the vegetable to the animal kingdom by expressing the hope that his botanical arithmetic "may possibly be applied at some future day with advantage to the different classes of Rotiferae in the animal creation."[18] A friend of Humboldt's went on to do precisely that, surveying the "wheel-animals" and all the other simple organisms then grouped under the general label of infusoria. Through comparative numerical studies, C. G.

Ehrenberg found that these active little animals were cosmopolitan if their usual habitat was marine and markedly localised if from freshwater sites.[19]

There were parallel studies in other branches of zoology. Karl Illiger, for instance, calculated the native countries of some 3,800 species of birds according to their frequency in various quarters of the globe—although this would perhaps have been more instructive if it had been done on a smaller scale.[20] Likewise, the French entomologist André Latreille showed which groups of insects increased in number toward the north pole and which toward the equator, confessing that it would be impossible to do for the geography of insects what Humboldt had done for the geography of plants.[21] Here Latreille was being unnecessarily modest, for it was entomologists in particular who found statistical or numerical accounts of their subject most rewarding. Entomology seemed to lend itself to the methods of botanical arithmetic, perhaps because insects are relatively numerous and thence amenable to statistical analysis, or, more possibly, because entomologists were usually well acquainted with botanical literature and were forced to consider plant distribution in relation to the life-style of insects. Either way, the transition from botanical to entomological topics was relatively easy, and botanical arithmetic soon passed from one field to the other.

So, for example, the Parisian entomologist Jean Théodore Lacordaire, a man of affairs, one of four brothers in public circles, and compiler of the insect section of the *Suites de Buffon*, had part of his *Introduction à l'Entomologie* translated into English and published at length in the *Edinburgh New Philosophical Journal* through 1839.[22] In every department of geographical distribution he used the statistical method for the computation of ratios and the tabular format for presentation of data.

Citing the botanical method, Lacordaire calculated the number of insect species per genus as Candolle, some twenty years earlier, had calculated the number of plants. He quoted figures that enabled him to conclude that although species may increase from north to south, genera are relatively more numerous in northerly regions. "From this it follows," he wrote, "that the absolute number of genera augments from the north to the south, since Europe has more than Siberia, and South America more than Europe; but it will be seen, at the same time, that this number does not increase in the same proportion as the species, but that it follows, on the contrary, an inverse progression."[23] Lacordaire was thus using the "simplest reckoning" to make fairly elaborate statements about the structure of insect populations on a global scale (figure 3).

Lacordaire's "Geographical Distribution" presented an interesting panorama of the variety of arithmetic devices naturalists were prepared to use. In it he gave figures to indicate the sparsity of "phyto-saprophagous" insects, which exist on decayed plant matter, compared with ordinary plant- or leaf-eating insects, and compared these again to dung-eating beetles in different regions of the globe (figure 4). Similarly, he listed twenty-two families of beetles and bugs and itemised the number of genera found in eight predetermined regions (fig-

Country.	Species.	Genera.	No. of Species, per Genus.
Siberia, ...	465	169	2.7
Europe, ...	5,677	715	7.9
North America, ...	2,403	541	4.4
South America, ...	8,112	1209	6.7
Africa, ...	2,942	674	4.3
New Holland, ...	320	162	2.

Figure 3 Lacordaire on the geographical distribution of insects. From T. Lacordaire, "On the geographical distribution of insects," *Edinburgh New Philosophical Journal* 27 (1838–39): 346.

	Number of Phytophagous Species.	Phyto-saprophagous Species are to the entire amount of Phytophaga,—
Lapland, . . .	529	: : 1 : 261
Sweden, . . .	1438	: : 1 : 219
England, . . .	2622	: : 1 : 180
France, . . .	3276	: : 1 : 188

Figure 4 Lacordaire on the proportion of leaf-eating insects. From T. Lacordaire, "On the geographical distribution of insects," *Edinburgh New Philosophical Journal* 28 (1839–40): 171.

ure 5). An immediate parallel with H. C. Watson's tables of altitudinal distribution springs to mind, and indeed, except for the different organisms employed, they are exactly the same. Lacordaire even gave the "proportion to 100 genera," just as Watson had given "Proportions" in his table (cf. figure 2).

And, again like Watson, Lacordaire presented a grand finale with close-packed columns of figures conveying the relative numerosity (by species) of all insect families known to science (figure 6). Each vertical column enclosed a hierarchical arrangement of families according to the number of species they presented in any one area. Reading across

Families.	Total of Genera.	No. of Genera proper to								Proportion to 100 Genera.	
		Regions 8.	Regions 7.	Regions 6.	Regions 5.	Regions 4.	Regions 3.	Regions 2.	Region 1.	Endemical.	Sporadical.
Carabidæ,	183	6	2	8	7	11	20	24	105	57	43
Hydrocanthares,	26	...	1	4	5	1	7	3	5	19	81
Brachelytres,	72	...	1	3	3	2	17	14	32	44	56
Sternoxes,	114	1	2	2	3	15	15	22	54	47	53
Malacodermes,	54	1	...	1	5	2	9	9	27	50	50
Teredyles,	31	1	...	...	2	4	10	3	11	35	65
Clavicornes,	63	1	...	4	5	7	7	11	28	44	56
Palpicornes,	11	1	...	2	1	...	2	4	1	9	91
Lamellicornes,	197	3	3	5	7	9	11	33	126	63	37
Malasomes,	124	...	1	1	...	2	14	22	84	67	33
Taxicornes,	54	...	...	...	2	6	6	8	32	59	41
Tenebrionites,	56	...	1	2	...	...	3	12	38	67	33
Helopidæ,	35	...	...	2	2	2	1	4	24	68	32
Trachelides,	26	...	..	2	2	1	3	6	12	46	54
Vesicants,	19	...	...	1	3	2	2	3	8	42	58
Stenelytres,	14	...	...	1	...	2	1	3	7	50	50
Curculionidæ,	350	1	3	1	12	17	17	59	240	68	32
Xylophages,	56	...	1	1	4	7	9	8	26	46	54
Longicornes,	348	1	1	1	3	12	17	47	266	76	24
Chrysomelidæ,	328	1	3	3	10	11	29	49	222	67	33
Trimeres,	38	...	1	2	1	4	2	6	22	57	43
Dimeres,	12	...	..	...	...	...	2	...	10	83	17
	2211	17	20	46	77	117	204	350	1380	62	38

Figure 5 Lacordaire's table of the number of genera appearing in one or more regions of the globe. Very few genera appeared in all eight regions (column one). The regions are named in figure 6. From T. Lacordaire, "On the geographical distribution of insects," *Edinburgh New Philosophical Journal* 28 (1839–40): 177.

the page gave one an idea of which families were most in evidence in the different parts of the world; the diligent naturalist might discern that the three northernmost landmasses of North America, Lapland with Siberia, and Asia all had the same family as their most plentiful insect—the *Carabidae*—whereas Africa and Australia possessed the lamellicorns in common, and Europe and India, *Curculionidae*. In this way, numerical methods illuminated problems of distribution that might never have been noticed otherwise.

Turning from entomology to marine invertebrates, both Edward Forbes and the American naturalist James Dana presented their results in equally significant tables. In 1843 Forbes produced a series of "conchological tables" to illustrate his *Report* on the mollusca and radiata of the Aegean seas to the British Association for the Advancement of Science.[24] In them he attempted to show the distribution of marine life according to depth in one or more of the eight zones into which he had divided the sea floor and, furthermore, to show the percentage of, say, the Celtic forms in each assemblage. James Dana did much the same thing with his succinct tables of polyp distribution,[25] one of which is reproduced here (figure 7). In 1847 he collected

North America.		South America, (including Mexico).		Africa.		Europe and Southern Russia.		Lapland and Siberia.		Asia.		Indian Archipelago.		New Holland.	
Families.	Species.	Families.	Species.	Families.	Species.	Families.	Species.	Families.	Species.	Families.	Species.	Families.	Species.	Families.	Spe.
Carabidæ .	389	Chrysomelidæ	1853	Lamellicornes	685	Curculionidæ	1145	Carabidæ .	182	Carabidæ .	194	Curculionidæ	238	Lamellicor.	70
Chrysomelidæ	375	Curculionidæ	1615	Carabidæ .	475	Carabidæ .	1090	Brachelytres	135	Lamellicornes	172	Chrysomelidæ	192	Curculioni.	65
Curculionidæ	285	Longicornes	947	Chrysomelidæ	387	Brachelytres	683	Chrysomelidæ	121	Chrysomelidæ	135	Lamellicornes	116	Chrysome.	60
Sternoxes .	211	Lamellicornes	905	Curculionidæ	314	Chrysomelidæ	610	Curculionidæ	117	Curculionidæ	130	Longicornes	105	Longicor.	35
Longicornes .	186	Carabidæ .	591	Melasomes .	312	Lamellicornes	388	Clavicornes .	82	Melasomes .	112	Carabidæ .	94	Sternoxes	32
Lamellicornes	172	Sternoxes .	451	Sternoxes .	213	Clavicornes .	380	Hydrocanthares	72	Sternoxes .	73	Trimeres .	78	Carabidæ	30
Malacodermes	130	Malacodermes	445	Longicornes	190	Sternoxes .	327	Longicornes .	58	Longicornes .	58	Sternoxes .	72	Helopiens	15
Brachelytres	129	Clavicornes .	192	Vesicants .	132	Longicornes	303	Sternoxes .	55	Vesicants .	50	Malacodermes	51	Malacoder.	10
Clavicornes .	117	Malasomes .	185	Malacodermes	103	Melasomes .	290	Lamellicornes	53	Clavicornes .	31	Clavicornes .	32	Melasomes	10
Trachelides .	88	Helopiens .	180	Clavicornes .	75	Malacodermes	270	Xylophages	47	Trimeres .	22	Taxicornes .	30	Trachelides	7
Xylophages .	85	Trimeres .	147	Xylophages .	56	Xylophages .	215	Trimeres .	38	Malacodermes	20	Tenebrionites	23	Hydrocantha.	6
Hydrocanthares	76	Xylophages	140	Trachelides .	50	Trimeres .	192	Malacodermes	37	Helopiens .	18	Helopiens .	18	Clavicornes	6
Teredyles .	61	Tenebrionites	132	Trimeres .	45	Hydrocanthares	156	Melasomes .	27	Hydrocanthares	17	Trachelides .	13	Tenebrioni.	6
Trimeres .	47	Taxicornes .	128	Helopiens .	43	Teredyles .	127	Taxicornes .	25	Teredyles .	14	Hydrocanthares	10	Stenelytres	5
Tenebrionites	40	Brachelytres	115	Hydrocanthares	38	Trachelides .	114	Vesicants .	22	Palpicornes .	11	Brachelytres .	10	Trimeres	5
Vesicants .	35	Vesicants .	88	Taxicornes .	34	Vesicants .	105	Palpicornes .	20	Tenebrionites	10	Melasomes .	7	Teredyles	4
Taxicornes .	30	Trachelides .	72	Tenebrionites	30	Taxicornes .	88	Trachelides .	16	Brachelytres .	9	Teredyles .	6	Brachelyt.	3
Palpicornes .	27	Hydrocanthares	65	Teredyles .	27	Helopiens .	80	Teredyles .	11	Taxicornes .	8	Xylophages .	5	Palpicornes	3
Helopiens .	25	Teredyles .	60	Brachelytres .	22	Palpicornes .	78	Stenelytres .	6	Trachelides .	8	Vesicants .	5	Taxicornes	3
Stenelytres, .	12	Palpicornes .	51	Stenelytres .	13	Pselaphidæ .	63	Tenebrionites .	5	Xylophages .	7	Palpicornes .	2	Vesicants	2
Melasomes .	7	Stenelytres .	20	Palpicornes .	12	Stenelytres .	57	Helopiens .	3	Stenelytres .	5	Stenelytres .	2	Xylophages	0
Pselaphidæ .	4	Pselaphidæ .	5	Pselaphidæ .	0	Tenebrionites	52	Pselaphidæ .	0	Pselaphidæ .	0	Pselaphidæ .	0	Pselaphidæ	0
	2531		3387		3229		6813		1137		1104		1109		375

Figure 6 Lacordaire's summary of insect distribution. From T. Lacordaire, "On the geographical distribution of insects," *Edinburgh New Philosophical Journal* 27 (1838–39): 349.

this set of figures and arranged them to display the limited geographical range of corals: each column represented the number of species confined to the region named at the head, so that consideration of (for instance) columns two and four would show that of fifty species found in the Pacific Ocean only four were also found in the East Indies or Indian Oceans. Like Lacordaire, Dana found that such tables allowed the presentation of fairly complex data in a readily comprehensible form. Tables ostensibly made it easier to compare and contrast; to convey subtle variations in the balance of this family of organisms with that family or of one region with another.

TABELLENSTATISTIK

Yet to modern eyes many of these numerical surveys seem somewhat pointless. The figures were rarely used to substantiate specific hypotheses, nor did they generate any important new questions about

	East Indies, Indian Ocean, or Red Sea.	Pacific Ocean.	West Indies.	Pacific, and the East Indies or Indian Ocean.	Extra-tropical.	Doubtful.	Total.
TRIBE ASTRÆACEA.							
Fam. Astræidæ, . .	37	50	29	4	3	16	139
Fungidæ, . .	14	29	6	6	0	10	65
TRIBE CARYOPHILLACEA.							
Fam. Caryophyllidæ,	13	7	9	2	13	5	49
Gemmiporidæ, .	4	5	1	2	0	2	14
TRIBE MADREPORACEA.							
Fam. Madreporidæ, .	30	42	4	8	1	7	92
Favositidæ,* .	14	15	5	3	0	4	41
Poritidæ† . .	5	14	6	2	0	1	28
	117	162	60	27	17	45	428

Figure 7 James Dana on the geographical distribution of coral polyps. From J. Dana, "On the geographical distribution of Zoophytes," *Edinburgh New Philosophical Journal* 43 (1847): 44.

geographical phenomena. At most, they were arranged as in Lacordaire's tables to illustrate long-established axioms of distribution, such as an increase in species number toward the equator or the simple fact that the relative incidence of certain families changes from region to region. No one asked why statistical methods were used or what the figures actually meant for natural history. It appears that the results of calculations such as Watson's on the relative proportions of plant families at different heights did not lead him—or anyone else—on to the construction of explanations for those facts. Watson, for instance, did not add any new concepts to his first essay on montane distribution during the thirty years or so after its publication, despite a continuing concern to refine and update the practical information given in successive volumes of his *Cybele Britannica*. In other words, his statistical work did not give him any new ideas.

This state of affairs was hardly what Humboldt had had in mind when he first thought of "botanical arithmetic." The wave of enthusiasm for distributional arithmetic obscured, and even obliterated, the greater theoretical aims put forward by Humboldt, Brown, and Candolle. Where the three of them had had an immediately visible and tantalising goal in the disclosure of laws of nature, their successors had nothing of the kind. They could not see beyond the numbers. Watson firmly told his contemporaries that the object of "geographical botany" was to gather the facts, not to fool around with theories:

> A first call on the phyto-geographer is to ascertain where the plants are now distributed, and in what manner and degree such present distribution accords with present geographical conditions. . . . The ingenious minds that take their delight in inventing causal hypotheses to account for these present seeming relations, through reference to supposed conditions of the past, will work more easily, and perhaps with more truthful results, if supplied with properly arranged data truly illustrating the present. They may then possibly find out, instead of only feign, phyto-geographical and phyto-geological histories.[26]

Here, with a passing swipe at Edward Forbes's "phyto-geological" theories (Forbes had appropriated Watson's scheme for the subdivision of the British flora to back his highly speculative account of the geological history of the British Isles),[27] Watson epitomised the aims of his distributional colleagues. For them the figures and the "facts" were the objects of research.

Echoing Watson's remarks, many British naturalists consciously and conscientiously restricted their investigations to numbers alone—to numbers bereft of any explanatory coda. There was a flurry of calculation for its own sake, generating a whole genre of purely numerical exercises in geographical distribution through the 1840s and 1850s. Later, this preoccupation with figures was derisively called *Tabellenstatistik*, after a similar episode in German political economy where, for a short period, theory was disregarded in favour of numerical data alone. If *Tabellenstatistik* had any meaning at all for biologists of these two decades, it meant that they ought to count and tabulate the phenomena of nature rather than seek to explain them. The tide of numbers reached its height with Alphonse de Candolle (son of Augustin), who declared that he loved figures as much as others detested them. He took the genre to its climax with his *Géographie botanique raisonnée*, published in 1855, after which this particular approach to biology declined dramatically. Perhaps not the least factor in its decline was the publication of Darwin's *Origin of Species* four years later.

More than just a curious interlude in nineteenth-century natural history, this intense preoccupation with figures deserves closer examination. It shows in more ways than one the manner in which naturalists struggled to cope with the changing content of their subject and with the intellectual demands of contemporary science. It reveals, moreover, the underlying models on which natural history and geographical distribution studies were based at that time.

First, although it may be hard to believe, there were strong emotional grounds for concentrating on figures alone. As many as 161,000 species unknown in 1750 could be found in collections in the year 1833. The pressures on naturalists to get them all described and properly integrated into current systems of thought must have been great. An unnamed species pains the tidy mind of a systematist, and a whole herbarium or cabinet of unidentified objects demands more time and effort than can necessarily be supplied. Each form could well be the subject of a monograph.

What then could be done with the ten thousand species of insect that Johann Christian Fabricius collected on his own behalf? In cases like this the arithmetical method was seized on with relief. It did not free naturalists from the task of identifying and naming their specimens, but it did give them the feeling that these facts were somehow being collated—synthesised by the mere activity of addition and tabulation. Here, a confusing jumble of data could be arranged into neat columns of numbers, perhaps under geographical headings, and certainly under family or order, giving everyone the satisfaction of doing something apparently useful.

When the facts were presented as numbers in this fashion there was a tangible feeling on the part of the author and the reader that the natural historian was in control of his subject. He appeared to know what he was doing, instead of being submerged in a morass of species. Because naturalists admired the strictly mathematical sciences, such an overwhelming use of numbers lent an inescapable appearance of authority to any given piece of work.

The last sentiment has, of course, occurred regularly in the history of biology, notably in the early development of fields such as ecology, where it was important to be seen to be doing something that others would call science. For nineteenth-century distribution studies, arithmetic served an important function by giving essentially descriptive work an air of precision, and hence of legitimacy. Such an air was a necessary prerequisite for any scientific enterprise in the nineteenth century, because it was an age when the demand for exactitude was at its height and exactitude was best served by the application of numbers. Nor were the sciences alone in this belief: in area after area spokesmen called for an accumulation of solid information, and the intellectual community laboured to respond in the appropriate fashion. In the larger context, figures and probabilities were even then changing the face of economics, life insurance, parliamentary legislation, medicine and medical policy, and other social issues that can be broadly grouped together under the term *public philanthropy.*[28] Elsewhere, Laplace and Quételet each in his own way had stimulated the development of modern statistics, which was rapidly acquiring all the institutional trappings of a formal discipline.[29] And at yet another level, figures were the source of great public interest and controversy—in the British Isles at least—in that the number of human inhabitants was the subject of fierce debate during the earlier years of the nineteenth century. More than any other contemporary topic, arguments over population brought simple arithmetic into the lives and homes of the people.[30] With this sort of background it is easy to see that mathematical procedures in the sciences were part of a generalised push toward numbers in all walks of life.

As for science itself, quantification was the order of the day. Queen Victoria's long reign saw the mathematically equipped scholars taking over the brunt of physical investigations, and familiarity with numbers became something of a byword for "proper" scientific procedure.[31]

Mathematical predictions led to exciting physical discoveries, like the existence of the planet Neptune; the science of heat was put into mathematical form by Fourier, the newer one of electromagnetism being similarly treated by Ampère; Young's empirical concept of the wave motion of light was given arithmetical form and content by Fresnel; Faraday's ideas were given mathematical shape by James Clerk Maxwell. Here, of course, the use of statistics was closely interwoven with both conscious and unconscious theoretical constructions: it is misleading to imply that a stress on numbers must necessarily lead to an avoidance of theory. But the public image of scientists was changing in directions that elevated mathematical ability to a prominence not enjoyed before. Figures of the past who had introduced calculations into astronomy—thereby making it queen of the sciences—were saluted as idols of a new age. Kepler and Newton enjoyed renewed attention in this respect, representing in their work the twin ideals of mathematical genius and a mass of synthesised empirical data.

The trend ran beyond the physical sciences. Botanical arithmetic was to natural history what mathematics was to the study of electromagnetism, heat, or light; it promised great achievements in the organisation and analysis of intractable data, and it successfully harnessed the growing enthusiasm for figures and numbers which was so much a part of nineteenth-century life. To botanists and zoologists of the period it must have seemed that a good round figure was worth a thousand words, and the application of numerical reasoning the essence of a scientific approach to nature. Tables and figures, ratios and proportions, therefore added weight and an aura of "science" to natural history publications. If nothing else, statistics in distribution studies was part of a larger movement towards the quantification of science.

THE DEMOGRAPHY OF SPECIES

There were other reasons, too, why *Tabellenstatistik*, unadulterated by any theory of distribution, had such a hold on the natural history of the period. These can best be explained by examining the model on which botanical arithmetic was based.

Tabellenstatistik was firmly rooted in an analogy with human population surveys, an analogy which natural historians made perfectly explicit in their writings. Hewett Cottrell Watson and his less famous contemporaries (such as C. C. Babington and J. H. Balfour) took care to draw a parallel between their surveys and those of the Department of Public Records, saying that they intended to count individual plant species as if they were the people of a kingdom. Watson likened his work on British plants to a census, with the important qualification that "in the former we count up only *species* or *habitats*," while in the latter it was individuals who were counted. "We cannot count," he wrote, "the individual plants of a species as we would count the people of a kingdom; and in an island so large as Britain, we cannot ascertain the number of special localities, unless it be for some few extremely local species. Their census must therefore be vaguely estimated by the experience of individual observers, who will never agree in their estimates; or it must be tried by some test which is adapted to show the number of *spaces* (not the number of precise *places*) in which each one occurs."[32]

Students of distribution evidently thought of themselves as proceeding along the same lines as a census, a point supported by the human demographic metaphors and imagery sprinkled through the writings of the period. Watson, as already mentioned, wrote of his *Cybele* as a great floral "census." He further divided the British Isles into "Provinces," "Counties," and "Vice-Counties" to indicate the areas in which specific plants could be found. Other authors used the words "nation," "metropolis," and "kingdom" to depict large-scale associations of organisms or their apparent demographic focus. Descriptions of "tribes" and "clans" are to be found with phrases like "the numbering of the populace" or "the counting up of species." Even the use of the term "statistics" implied a recognition of the transfer of this element from human society to natural history: Watson took pleasure in calling himself a "botanical statist" and endeavoured to produce "botanico-statistical comparisons." In 1852, in a *Class Book of Botany*, John Hutton Balfour finally codified the "statistics of vegetation" as the "counting up of the plants of every kingdom," thereby fixing the demographic metaphor as something that had pervaded all levels of botanical endeavour.

The geologist Charles Lyell—by no means a proponent of mere *Tabellenstatistik*—produced one of the more interesting examples of this analogy between the natural world and human society. He likened himself to a demographic commissioner who visited a number of regions in turn to compile a register of their fossil inhabitants. The geologist-commissioner took an "exact account of the number, names and individual peculiarities of all the inhabitants" and left in each district a "register containing a record of this information." Lyell explained: "If after the completion of one census, another is immediately made after the same plan, and then another, there will, at last, be a series of statistical documents in each province. When these are arranged in chronological order, the contents of those which stand next to each other will differ according to the length of the intervals of time between the taking of each census."[33] Clearly, Lyell saw the metaphor of a census as a reflection of his own views on the nature of the fossil record: that changes in the composition of living associations and the geological structure of the earth were brought about gradually, even though the fossil record and stratigraphical table were seemingly built on discontinuities. It was the taking of a "census" that made breaks in the chain of continuity and gave a false impression of periodisation in the history of the living world. Lyell's imagery and his vocabulary were drawn from human affairs; this was the language of political arithmetic.

The parallel between human and nonhuman demography ran even deeper than these verbal associations and similes. The aim of British censuses in the first half of the nineteenth century was—obviously enough—to count the number of people living in the British Isles. Unfortunately, methods then in use were hopelessly inadequate for the task. A constant succession of parliamentary bills changing the procedure of assessment led to different people being counted every time; although censuses had been taken at ten-year intervals since 1801, nothing was standardised, rigorously checked, or applied to more than a limited group of landowners, tenants, and taxpayers until 1841, when reorganisation began. Before that time, as Lyell so aptly pointed out, even the process of enumeration was chaotic. Temporary government officials (often poorly educated) visited each village, town, or street in turn, doing no more than list the number of adults in every house on their route. Inaccuracies were rife, and produced discouragingly distorted results. Politicians were therefore unable or unwilling to draw conclusions from such unreliable information, and important Acts designed for the relief of the poor or the increased taxation of richer Britons foundered on a lack of quantitative guidelines. The cry went up for facts, numbers, and tables.

In the natural history sciences, scholars were in the same predicament as politicians—although, of course, their species could not complain or answer back. They were faced with a multitude of new and disturbing problems, some arising directly from the influx of specimens to European museums and others out of the phenomenal abundance of life that was becoming increasingly apparent. They too needed to know the demographic particulars of their subject. Unfortunately, their acts of Parliament—that is, their theories of creation, classification schemes, and explanations of the causes for distribution—were confusing the issue instead of clarifying it. Here, as in demography, the cry went up for facts, numbers, and tables.

And the biological community laboured hard to satisfy the need. Their numerical surveys were designed specifically to imitate a census. The purpose of the exercise rested in the figures, not in the conclusions which might be drawn from them. Like a census, the results might indeed reveal something unknown or unexpected about the natural world, just as the total population of Great Britain was revealed through decennial surveys. But it was not the job of demographers to integrate these facts into current policy; that was for Parliament to decide. In a similar vein, natural historians relieved themselves of the burden of explaining their numerical findings, taking shelter instead behind the demographic model as justification for their actions. "Botanical statists" saw themselves as the fact-finders, the demographic commissioners of the natural world. They saw themselves as accumulating the raw material for some long-awaited Newton of the biological sciences, a man who would discover great universal truths in their work. Meanwhile, the thing to do was to count.

THE NEW BOTANICAL GEOGRAPHY

This last view sounds very Baconian and, sure enough, the practitioners of distributional arithmetic believed that they were following the method prescribed by the Lord Chancellor. But, as several historians have pointed out, it was a peculiarly nineteenth-century Baconianism, so divorced from the ideas of the seventeenth century that Susan Cannon has recently suggested that it should be named after Humboldt, rather than Bacon.[34] Be this as it may, the interesting point is that neo-Baconian (or quasi-Baconian) principles were in vogue, and natural historians thought they were following them to the letter. In 1847 Watson prefaced his *Cybele* with a stirring epigram on the practice of science, citing a line or two from John Herschel, the most influential spokesman for the sciences in the first half of the century.[35] Although Herschel argued that the best kind of scientific statement was a quantitative one and glorified Newton's inverse square law as the pinnacle of modern science, Watson found more significance in Herschel's belief that science was "the digestion and arrangement of facts which, when so arranged, will tend to show causes and their effects and the laws of nature."[36] He interpreted Herschel as saying that real science was the product of orderly and methodical arrangements of facts—a perfect definition of the activities of distributionists at the time. So, for Watson and for other self-consciously scientific botanists and zoologists, the statistics of distribution was an effective demonstration that natural history was keeping up with the push toward mathematics in other disciplines. The technique made the study of distribution appear to be a truly "scientific" enterprise.

Certainly, this was how the more thoughtful naturalists saw the question. In 1845 Joseph Hooker, then on the threshold of a career as one of Europe's most noted botanists, exclaimed that numbers were essential for the advance of distributional studies: "We must know the absolute numerical amount of peculiar species. . . . No-one has attempted this, all seem to dread making Botanical Geography too exact a science; they find it far easier to speculate than to employ the inductive process. The first step to tracing the progress of the creation of vegetation is to know the proportions in which the groups appear in different localities, and more particularly, the relation which exists between the floras of the localities, a relation which must be expressed in numbers to be at all tangible."[37] Hooker obviously felt that numbers were basic to botanical science, as did his correspondent, Darwin, who was to echo Hooker's view in regretting that zoologists "seldom go into strict and disagreeable arithmetic like you botanists so wisely do."[38]

Hooker went on to publish authoritative data on the vegetation of Antarctica, Australia, and New Zealand, the results of his four-year voyage (1839–1843) as surgeon-naturalist with James Ross. In a series of heavily didactic essays introducing separate catalogues of plants, Hooker outlined the principles of systematic and geographical botany for the benefit of colonial scholars, emphasising in turn the high degree of accuracy required of any scientist, the difficulties attendant on classification, the danger of adding new species to over-large inventories, the likelihood of species being more variable and widely spread than was usually supposed, and, above all, the necessity for proven facts unadulterated by speculation. He complained of the "rashness," "vagaries," and "recklessness" of other botanists who crowded textbooks with "defective definitions" and "erroneous classifications."[39] In addition, he took exception to "hasty generalisation or loose speculation."[40] Hooker claimed that, contrary to other workers in the field, he was "fettered by no professed opinion to support, and had no inculcated theory to eradicate."[41] Numbers, for him, were the best sort of fact attainable, since they alone were devoid of theory.

Here, of course, it is proper to ask to what extent Hooker really believed in these ideals and how much of his commentary was dictated by the conventions of the day. The cynic might claim that Hooker was really only justifying the time he spent on botany by making it out to be rather more difficult than the stereotypic image would suggest. He certainly called it a "difficult and exacting subject." Perhaps there is truth in such an interpretation. But it seems more probable that Hooker was genuinely devoted to nineteenth-century ideas about the nature of science: genuinely committed to the pursuit of truth, to personal and professional integrity, and to the dissemination of knowledge in the face of social and topographical barriers.

So, in the intellectual climate of the second quarter of the nineteenth century, where figures and numbers commanded attention at every turn, it seems only natural that Hooker should feel strongly about the need for such numbers in botanical geography. These, and these alone, would make the subject an exact science. Hooker's feelings on the matter were eclipsed, however, by the thundering endorsement of numerical botany delivered by his colleague Alphonse de Candolle in 1855. Son of the great Genevan botanist and humanist Augustin de Candolle and noted systematist in his own right, Alphonse was the real spokesman for botanical geography in this era. He extended many of the investigations begun by his father and wrote of the benefits of mathematical analysis and numerical comparisons in the nascent field of phytogeography. Alphonse de Candolle believed the use of statistics was a mark of maturity in a science, and he was anxious that botany should not, so to speak, miss its coming of age:

> Il y a dans chaque science, art ou objet d'étude, une partie statistique, soit numérique. On la retrouve en agriculture, en médecine, dans toutes les branches de l'administration, dans les sciences physiques, naturelles, et jusque dans les sciences morales. Elle occupe une très grande place dans la géographie botanique. Pour moi, j'en conviens, j'aime les chiffres autant que d'autres les détestent.[42]

And elaborating on these remarks in his important *Géographie botanique raisonnée* (1855), Candolle *fils* confessed that he was dedicated to the principle of arithmetical exactitude in distribution studies. He intended to illustrate the arrangement of the vegetable kingdom through the use of proportions and the decimal system. Furthermore, he took a leaf out of Charles Lyell's book and claimed that he was setting out the "Principles" of a new science—in this case the study of plant geography, rather than geology. Like Lyell, he brought a formidable analytic talent to the study of natural phenomena; like Lyell, he too had a single unifying theme on which the new discipline ought to be based, although this seems to have been more a naïve faith in precision than a fully articulated scheme as found in the *Principles of Geology*. If the facts were right, asserted Candolle, and if they were ordered and collated with all due caution, then inescapable truths of nature could be derived from them. Many facts, of course, could not be treated arithmetically, but where such a method was appropriate it was essential to use it. Candolle considered that his own extensive use of statistics established a precedent in natural history, and had led to the formulation of generalisations about plant distribution which brought order out of chaos. He felt justified in calling his treatise géographie botanique *raisonnée*:

> Cette appréciation des méthodes, ainsi que la discussion des faits et la tendance à rechercher les causes des phénomènes, m'ont paru établir une assez grande différence entre mon travail et la plupart des traités de géographie botanique, dans lesquels on se contente de décrire des pays ou de citer des faits.[43]

Here, perhaps, Candolle was succumbing to the temptation to make a general condemnation of his precursors and contemporaries, just as Lyell some twenty years earlier had distorted the aims and achievements of geological theorisers of previous generations. Candolle may indeed have wished to rewrite history like Lyell, whose motives have been brought out in all their complexity by M. J. S. Rudwick and Roy Porter in recent papers.[44] Yet in many ways he had accurately assessed his own position in the history of biology, for of those naturalists who used statistical methods or attempted to generalise about floral relationships and historical causes, few had done so with such care, consistency, and scope as he had. The *Géographie botanique raisonnée* was undoubtedly a great contribution to its field and stood with Humboldt's *Voyage aux régions équinoxiales* as one of the classics of phytogeographical literature.

But the point is not so much Candolle's unsubtle self-advertisement as the fact that he saw himself as giving shape and method to an apparently amorphous body of information. He outlined the techniques of analysis, the modes of reasoning and their limitations, and the extent to which conclusions could be drawn, all of which were extensively substantiated by his own detailed example. And this example was propelled by what can only be called statistical monomania: "Elle occupe une très grande place dans la géographie botanique," he had truthfully written in his preface, but a naturalist may be forgiven for wondering—after reading the two substantial volumes that followed these words—how Candolle allowed plants into the book at all. Candolle deliberately treated the "facts" of distribution as data to be processed and tabulated in the search for higher laws, and each plant or specimen was therefore only useful as a single unit in the mass of raw material that had to be reduced into a semblance of order. Statistics, he thought, were to transform plant geography into a genuine science.

With the work of the younger Candolle, botanical arithmetic returned from the atheoretical wilderness, so to speak, and once again became the means by which botanists could work toward fundamental truths of nature. Candolle was anxious to deal with all aspects of plant geography and described in succinct fashion what was—and was not—known to scientists working in this field; he was also anxious to introduce new generalisations and to outline the most promising or the most problematic areas in which to conduct future research. These aims were realised in the *Géographie botanique raisonnée* and the book consequently bears some importance for the subsequent development of phytogeography, in that it served as a focal point for contemporary workers. Candolle redefined botanical geography as the search for "des idées génerales," ideas based on carefully collated information. He presented detailed arguments for the variety of factors that influence and limit the ranges of individual plants and botanical communities, and ordered them in such a manner that most of the empirical data then available was set out in a logical fashion. For this reason, even his critics (Joseph Hooker was one) had to admire and congratulate him. The extent of Darwin's debt to Candolle, for example, is only now coming to be recognised. Besides an interest in Candolle's views on species and on the origins of island floras, Darwin took note of an overwhelming number of Candolle's descriptions, of, say, social plants, many of which were scribbled in the margins of his personal copy of the book. For Darwin, as for other naturalists, the work was indeed *raisonnée*.

More significantly perhaps, Alphonse de Candolle had demonstrated that statistics, on the one hand, and theory, on the other, were commensurable; he showed naturalists the way in which figures could be used to vindicate or elaborate upon hypotheses designed to *explain* geographical distribution. *Tabellenstatistik* was superseded by a new botanical geography where fact and theory were reunited, on paper, at least, if not in practice. The aims of earlier botanical "statists" had given way to those who wished to make the subject a "science," and the model of a demographic census had been transcended by a constellation of ideas relating to the mathematical and physical branches of scientific endeavour. Where at first it had seemed eminently "scientific" to collect and collate data in imitation of human population surveys, it was now thought that phytogeography could only be a science if the facts were actually used to substantiate theory. Returning to the ideas expressed by his father, Candolle understood phytogeography as an inquiry of the highest philosophical interest, replete with theoretical conundrums and full of meaning for the deeper knowledge of nature.

But the theories Alphonse wished to examine in 1855 were very different from those advanced by Augustin de Candolle in 1820. The general tenor of the life sciences had changed dramatically in the thirty-five years that separated the publications of father and son. The new science of geology had been constructed.

FOUR

A Science of Processes

Interest in animal and plant distribution was not confined to systematic botanists and zoologists alone. During the first half of the nineteenth century geographical phenomena also attracted the attention of their more geologically minded colleagues, who used the subject to extend the scope of palaeontological investigations and to further their understanding of the history of living beings. As a result, a whole new dimension was added to research in geographical distribution: quite literally, the concept of time was added to that of arrangement in space, allowing the development of a refined and truly historical appreciation of the modern world.

But the transformation from "topographical" to "historical" studies was not as clear-cut as it might seem. Several botanists, Hewett Cottrell Watson among them, were not sure that they wanted to include the past in their analyses of the present, and said as much. Others, led by Alphonse de Candolle, adopted the historical approach with enthusiasm but wondered how it could be utilised to answer specific problems. But many geologists and palaeontologists were unmoved by the subtleties of topographical inquiry and preferred to wait for the formulation of a set of rules about distribution that could be applied to the fossil record; they hoped these would supplement the explanations already in vogue. A handful of naturalists, on the other hand, were as anxious to understand the present as the past. Charles Lyell, Edward Forbes, and Charles Darwin attempted to link ancient and modern distribution patterns together as a real sequence over time. The introduction of "history" into the study of animal and plant distribution was therefore a fairly haphazard affair which owed more to personal proclivities and styles of scientific investigation than to any conscious effort to expand the content or improve the methods of the subject. Nevertheless, there seem to have been two distinct lines of "historical" inquiry during the first fifty years of the nineteenth century.

First, simple empirical statements about modern topographical arrangements were applied to the fossil record. This helped geologists and other naturalists to substantiate theories relating to the internal heat of the globe and apparent advancement of the organic world through successive epochs.

The second line of inquiry was born out of the first and, with the advent of evolutionary theory, superseded it later in the century. Inspired by Charles Lyell, a few naturalists realised that such ready "borrowing" of facts and theories from topographical studies was undesirable, in view of the fluid state of that subject. There was no recognisable consensus in the phytogeographical or zoogeographical sciences. Instead, they aimed to set up an interpretative substructure of their own that would combine both past and present patterns of distribution. In their hands, geographical distribution emerged as a study of organisms in space *and* time.

A DIRECTIONAL HISTORY OF THE EARTH

Under the influence of Buffon, many geologists of the early nineteenth century believed that the earth was undergoing some sort of unilinear and nonreversible change. Men such as Georges Cuvier, William Buckland, Alexandre and Adolphe Brongniart, and Adam Sedgwick thought that since its formation the earth had experienced a progressive and directional change through a succession of geological epochs and that its living beings had progressed from simple to more complex bodies in accordance with world conditions. As Peter Bowler has pointed out, these geologists accepted the notion that the apparent "advancement" of life through the ages was directly caused by changes in the environment, not through any Lamarckian idea of biological change matching altered external conditions, but through an assumption that each set of beings was created to suit its surroundings.[1] As the earth developed from an original primal state—usually postulated as one that was very hot, and in later years as one with a high concentration of either atmospheric water or carbon dioxide—to the conditions of geologically recent periods, so life itself changed to accommodate a series of different factors.

Since the geological structure and events of any one age could be independently deduced from the nature of its deposits, the fossil record was at first understood to be little more than an effect, or biological offshoot, of physical change. This view was accurately reflected in the notion, common to Cuvier and Giraud-Soulavie in France and William Smith in England, that fossils were essentially indicators of the different strata in which they were found. Later the situation was reversed, with palaeontological evidence taking precedence over the mineralogical or lithological constitution of rocks. Resolving the question posed by his predecessor as president of the Geological Society of London, Adam Sedgwick emphasised in 1831 that zoological characters were the best guide for identifying beds.[2] Only a few years later fossils had become every geologist's first choice. Roderick Murchison's greatest achievement was to link Old Red Sandstone with the Devon greywacke—two very different deposits—by finding their fossil contents mingled together in a third deposit in Russia. In this case the organic remains proved more significant than the rocks themselves. And by 1841 the traditional terminology of "Primary" and "Secondary" periods had been replaced by "Palaeozoic" and "Mesozoic," signifying that faunal complexes characterised the major eras more effectively than inorganic or purely chronological factors. Accordingly, emphasis shifted from using fossils as indicators of the history of the earth to using earth-history as a guide to the development of living beings. Either way, the links between geology and the organic realm had become firmly established. This particular understanding of life can be called "directionalist," since the progress of complexity as seen in the fossil record depended ultimately on the successive changes of the crust of the earth and not on any internal property of "improvement."

Cuvier's researches were essential to theories of the "directional" history of the earth. His work was based on an interdependence between the physical conditions of various geological epochs (as inferred from the rocks) and the living beings supported therein. In effect, this close relationship was an extension of Cuvier's belief in the unity of the organism, where the correlation of parts meant that the structure and probable physiology of an animal could be reconstructed from a single bone. "Every organised creature forms a whole, a unique and closed system, whose parts mutually correspond to one another and concur toward the same definite action by a reciprocal reaction," he explained in the *Discours sur les révolutions de la surface du globe* (1812).[3] The system was so closely knit that if one part of the body changed then all other parts would have to change also. It would be impossible, however, for any such alterations to take place without jeopardising the existence of the animal concerned. For this reason, Cuvier was convinced of the fixity of species. By analogy with the correlation of parts, he saw the fauna of the earth as part of a closed system with the environment. Each species was so well designed for its own peculiar conditions that it simply could not live under any others. If the earth's surface altered, all forms of life had to be extinguished. Given the belief that the earth had indeed changed from a primitive chaotic state to the conditions of today, it is not difficult to see how Cuvier hit upon the idea that sets of organisms had been periodically swept away to be replaced by others which approximated more closely to modern species.

The suggestion of a serial replacement of fossil groups, each more advanced than the last, was first expounded in Cuvier and Alexandre Brongniart's work on the Paris basin.[4] This masterpiece of stratigraphical analysis did much to confirm a "directionalist" point of view, for it presented striking evidence of the differentiation of formations on the basis of their fossil content, which varied according to the conditions under which the deposits were laid down. Marine and freshwater molluscs came and went according to the periodic influxes of Tertiary seas over the area. Further work at the gypsum quarries of Montmartre and on specimens from other sites revealed a number of distinct faunas of varying antiquity. Cuvier's findings were published in his influential *Recherches sur les ossemens fossiles des quadrupèdes* (1812), prefixed by an important essay (also issued separately as the *Discours sur les révolutions de la surface du globe*), in which he described a sequence of animal forms that were increasingly different from those of the present day. In superficial deposits he found the remains of organisms closely resembling their modern counterparts, such as the Siberian mammoth. Pressing deeper into the Tertiary, he reconstructed the *Palaeotherium*, unknown among living animals but obviously allied to both rhinoceros and tapir. In Secondary formations he could find no mammals; bones from the chalk of Maestricht and from the older beds of Thuringia represented faunas in which reptiles and fish predominated. Below these, more primitive fish and invertebrate remains were found. Even on these limited grounds it was clear that the animal world had undergone some serial "advancement" through time.

Cuvier's research and the "directionalist" thesis it conveyed was the starting point—and the model—for much of the palaeontological work of following decades. The most important development, so far as the history of biogeography is concerned, was the fusion of this directional hypothesis with a second line of argument over the internal heat of the earth.

This second theory suggested that the earth had once been a molten ball of fire which subsequently cooled and condensed, and in particular that heat stored at the centre of the earth was the source of all tectonic and volcanic phenomena. It accorded well with Laplace's nebular hypothesis for the origin of the planetary system, first proposed in 1769 and frequently expanded thereafter, in which it was argued that planets had condensed out of the rotating nebulous atmosphere of the sun following the ordinary operations of physics and chemistry. William Herschel's sidereal astronomy had given renewed weight to these ideas in the first years of the nineteenth century, and they were further supported by John Herschel's work of the 1820s and 1830s. The nebular hypothesis was widely accepted by the scientific community, thereby increasing support for the notion of geological heat.

Empirical evidence came mainly from the researches of physicists and geologists. The idea of a central heat gave rise to many explanations for movements of the earth's crust. Hutton's account of the consolidation and elevation of strata by heat, and the volcanic origin of Primary rocks, both were deeply rooted in this concept, and similar views were made known to a wider public by John Playfair in his 1802 *Illustrations of the Huttonian theory of the earth.*[5] The well-known debate between "Neptunists" and "Plutonists," Wernerians and Huttonians, took much of its tone from the same belief. "Neptunists," as the label implied, thought all great events in the history of the earth originated in water action; such ideas were most famously developed by Abraham Gottlob Werner and his followers. Werner believed that crystalline rocks such as granite had first solidified out of the original universal chaotic ocean. British Plutonists, on the other hand, such as Sir James Hall and Thomas Beddoes propounded the view that all crystalline and Primary rocks were of igneous origin and so derived from volcanic activity. The question whether whins—thin, hard layers of rock—and dykes were genuine intrusions of laval material was one extension of this controversial issue. Much further into the century, Cuvier's "revolutions" of the globe were linked by his pupil Élie de Beaumont to periods of intense mountain building caused by outbursts of heat-related activity. The forces that contributed to these upheavals seemed to be greater in earlier times, giving credence to the idea of a gradually decreasing source of power. And in 1820 the theory of a central heat was virtually confirmed—at least for nineteenth-century scientists—through the calculations of the mathematical physicist, J. B. Fourier. He found that geothermal gradients observed in European mineshafts were analogous to the heat flow of a cooling sphere and concluded that there was a residual heat at the centre of the earth.[6]

So by the 1820s, the astronomical theory of an earth intensely hot at its formation had come to be widely accepted. Allied to the evidence of residual heat in the present day and to the hypotheses of Cuvier, this belief led to the inescapable conclusion that the earth had cooled down over time. On this hypothesis most of the geological work of the following decades was based.

THE COOLING EARTH AND PALAEOBOTANY

Some of the strongest palaeontological support for a progressively cooling earth came not from zoological research but from the study of fossil floras. Fossil plants had been noted and displayed in collectors' cabinets since the time of Edward Lhuyd (1660–1709) and J. J. Scheuchzer (1672–1733) and were even catalogued by Scheuchzer in his *Herbarium Diluvianum* of 1709. But they were not really absorbed into geological thought until the issue of the heat of the earth arose in the closing years of the eighteenth century. Antoine de Jussieu, the first of the long line of distinguished botanists to bear that name, seems to have been the first naturalist to observe that plants found in coal deposits around Lyon were not typical of European—or any temperate—regions, but were more suited to the hot and humid climates of the equator. In 1718 he pointed this out to the Académie Royale des Sciences in a memoir, "Examen des causes des impressions de plantes marquées sur certaines pierres des environs de St. Chaumont dans le Lyonnais." The same fact was noted in 1801 by the German palaeontologist Baron Schlotheim who had studied at Freiberg under Werner and, for most of his life, superintended the ducal collections at Gotha.[7] By this time, mostly through Cuvier's efforts, it was becoming acceptable to suggest that fossil forms that differed greatly from existing organisms had become extinct. So Schlotheim asserted that the plants found in bituminous schists of Thuringia not only belonged to a warmer climate, but were extinct as well.

Despite only a handful of specimens (about 500 species compared with over 50,000 living plants known to botanists of the period,)[8] the study of palaeobotany soon came into its own with the investigations of Adolphe Brongniart, the son of Cuvier's partner. He recognised that since plants are more dependent on their surroundings than animals they should be of great value for determining the course of the earth's development. He surveyed the plant fossils of various geological periods, identifying and naming more species than anyone had done before, in order to reconstruct the physical history of the globe according to the Cuvierian model. This work was published as the *Histoire des végétaux fossiles* in 1828, accompanied by a short *Prodrome* in which he put forward theoretical views of great significance. The book was, in his view, an "exposition chronologique des périodes de végétation et des flores diverses qui se sont succédé a la surface de la terre."

Each of Brongniart's successive floras was different from the next, just as Cuvier's faunas had been. Together they formed a sequence of plant life that changed in character with every geological epoch, becoming more like the present state of things as time progressed. Individually, they were characterised by readily identifiable features. These features were not, for Brongniart, exclusively qualitative and dependent on the extinction or introduction of particular species at each "change-over," but included differences of a quantitative nature. He noted the relative proportions of the major plant groups such as mosses, ferns, cycads, and flowering plants for each epoch, and saw that these shifted gradually toward the proportions held in the present day.

Therefore, "sous le point de vue de la proportion numérique des espèces des différentes classes," he came to the conclusion that the Carboniferous flora was dominated by ferns and their allies (five-sixths of the total), whereas these vascular cryptogams make up barely one-thirtieth of modern vegetation. Alternatively, dicotyledonous plants, which make up three-fifths of today's flora, did not even exist in the Carboniferous epoch.[9] The preponderance of ferns and lack of any advanced plants in this period signalled two things to Brongniart: first, the environment, having higher temperatures and a moister atmosphere than today, must have been eminently suitable for cryptogams; and second, the flora was patently more primitive then than now. After successive calculations of this sort—obviously the same as the botanical arithmetic discussed in chapter 3—he concluded that the history of plants could be divided into four great *périodes*, each of which could be characterised by the proportions of the main classes. Each one was more "perfect" than the last, in that it approached the ratios of today and was marked by the introduction of more and more advanced forms of life. Brongniart concluded his *Prodrome* to the *Histoire des végé-*

taux fossiles with a small table illustrating these results, in which the number of species in different classes was listed in each of his four periods and compared with the number today. The dividing lines between *périodes* were, with only a little difficulty, collated with Cuvier's major shifts in the animal fossil record. He proudly remarked that these facts seemed to confirm that life was indeed created in a series of units according to the "révolutions de notre globe."[10]

Alphonse de Candolle followed Brongniart's interpretation of the fossil record almost to the letter. He added to and revised the lists of species and, at one time, divided the fossil flora into fourteen epochs instead of four, but he recognised with Brongniart that the "vegetable kingdom appears to have been gradually becoming more perfect."[11] Like Brongniart, he too described the physical conditions that must have accompanied particular species. In an 1834 article, "On the History of Fossil Vegetables," he gave an amended version of Brongniart's table to show the number of species from certain botanical classes appearing in each of the four geological eras which Brongniart had discerned. And again like Brongniart, Candolle used this table, and further figures of the same nature, to support the thesis that the history of plant life was one of sequential floras each differing considerably from the last, making a series progressively approaching the ratios of today's vegetation.[12]

Since so much of this interpretation depended on the assumption that a predominance of ferns and their allies indicated a hot and humid climate, the younger Candolle devoted most of his attention to a close examination of Carboniferous fossils, and, in particular, to the plants that made up coal deposits. Here, instead of calculating the relative proportions of classes through time, he proceeded just as if he had a problem flora of the present day and worked out the ratio of classes to each other and as a percentage of the total. Through these proportions he intended to discover exactly what climatic conditions allowed coal plants to flourish. The group he called *Aethogameae*, consisting of arborescent ferns, lycopodia, and equisetales, was seen to form upwards of five-sixths of the vegetation.[13] Statistics of this kind suggested either that the Carboniferous climate had been warm, moist, and equable or that coal plants had grown on innumerable, small oceanic islands. Because the lower cryptogams were "wholly wanting at this epoch," or at least were "very rare, as none have hitherto been discovered," Candolle reasoned that the plants living at that time could not have been oceanic. There were no seaweeds in coal deposits as far as he knew.

This reasoning provided him with useful ammunition for the current debate over a possible marine or brackish origin for coal. More important, it also confirmed the hypothesis of much higher surface temperatures in the past than in the present.

Candolle's arguments illustrate the highly abstract arguments fossilised remains were supposed to verify and the depth of meaning that could be read into a few numerical and taxonomic statements. These theories were not, so to speak, erected on organic evidence; rather the opposite. Fossils were thrown in as supplementary data for hypotheses that had already been conceived and that had other geological evidence in their favour. Both Brongniart and Candolle used botanical geography and systematics to substantiate ideas already circulating in geological society. They employed palaeobotany and the data of modern distribution to vindicate a particular interpretation of the earth's history.

BIOLOGICAL PROVINCES IN THE PAST

Adolphe Brongniart and Alphonse de Candolle both believed that life first appeared as a single primitive population evenly distributed over the entire surface of the globe. Gradually this uniform population fragmented into many diverse groups of species. As the surface of the earth itself had changed from conditions that were uniform, typified by a hot and stable climate, to those which showed great diversity in the distribution of land and sea and varying temperatures according to latitude, so, Brongniart and Candolle assumed, had the animal and plant kingdoms. As far as phytogeography was concerned, this view depended on the absence of biological provinces during the first geological periods that supported living beings. For Brongniart it was only in geologically recent periods that botanical provinces materialised out of the uniformity which he thought characterised earlier floras. Using Lyell's freshly coined names for the various subdivisions of the Tertiary, he pointed out that the first outlines of regional associations did not appear until the Eocene (the oldest formation of the Tertiary age). And even here, Eocene floras were not the same as those of today. They were spread more widely and included more families: this tropical flora, characterised by palms, algae, marine monocotyledons, and "extra-European" forms, extended from the lower latitudes to points as far north as the Parisian and London basins. In the same vein, Candolle, writing in 1834 on the Carboniferous flora, reviewed the evidence for a "universal" flora. The facts seem to prove, he claimed, "that at this epoch there was a greater uniformity of the vegetation on the surface of the earth, than exists in the times in which we are now living."[14]

The suggestion of a worldwide extension of early floras provoked an assortment of comments. In 1829, John Fleming correctly asserted that all the evidence for a former heat in Europe was precariously based on the identity of species from distant locations. Perhaps, he suggested, these species were only generically similar. The assumption rested entirely on analogy with the present day. He went on to note that a similarity of structure between fossilised leaves, stems, cones, and roots was of little use in discerning the habits, geographical range, or physiology of once-living species. Only the latter properties were a valid base for inference about previous climates.[15] Fleming's argument was soon rebutted by Alexander Murray, who proposed the opposite view that the anatomical structure of today's plants was indeed characteristic of the climates in which they lived.[16] Fossils related to existing tropical species must themselves have grown in a tropical climate.

The debate, such as it was, was short-lived. The idea of a single aboriginal population of species soon predominated. By 1830 an anonymous contributor to the *Edinburgh New Philosophical Journal* claimed that a "uniformity of climate prevailed over the earth prior to the time of the Deluge." He continued:

> It appears from the observations of geologists, that, during the earlier periods of the earth's formation, there did not exist, among the then created animals and vegetables, that kind of geographical distribution which characterises the organised beings of our time.[17]

His words reflected a growing feeling that animal remains conformed to the scheme laid down for plants. The evidence from Carboniferous floras was at that time supplemented by the striking case of Tertiary elephants. The Siberian mammoth and the Ohio mastodon had been known since the 1770s and were frequently conflated into one species (or one genus) with the living elephants, until Cuvier separated them in his famous memoir of 1799 "Sur les espèces d'elephans vivantes et fossiles."[18] Nevertheless, these extinct species were plainly elephants: analogy with living members of that group suggested that they had lived in warm climates, a point further substantiated by their great size (then thought to be characteristic of tropical species) and evident feeding requirements. Since mammoth and mastodon were closely related, warm temperatures were thought to have extended through the higher latitudes of Europe and North America; the New and the Old Worlds of the northern hemisphere must therefore have been under a uniform climate. These ideas were apparently confirmed in 1832, by Joseph Pentland's (mistaken) identification of giant bones from Australia as another species of proboscidean.[19] Although privately disagreeing with the notion of previous faunas spread worldwide under a much warmer climate, Pentland pointed out that this discovery tallied with the theory of a wide geographical range for Tertiary elephants.

But the editor of the *Edinburgh New Philosophical Journal* strongly disapproved. Robert Jameson added his own thoughts to the debate in 1832. As a staunch defender of Wernerian geology he was no supporter of theories that depended for their validity on the central heat of the earth. He criticised the characterisation of whole epochs by a "few miserable plant fossils." The change in relative numbers of cryptogams, gymnosperms, and angiosperms could just as easily be the result of irregular fossilisation or inadequate taxonomic identification.[20] This last complaint was perfectly justified in the light of contemporary palaeobotany. In some cases specimens were so poor that they could not even be assigned to a family—the most workable unit for botanists. Only a few pages beyond Jameson's comments in the 1832 issue of the Edinburgh journal, John Lindley demonstrated that very point. For the purposes of stratigraphy, identification of the major characters was all that geologists required; which was just as well, said Lindley, because these were all that the internal structure of the stem or the veining of leaves could tell one. There was simply no detailed evidence to back sweeping generalisations about former climates.[21]

The controversy over the uniformity of fossil floras and faunas forced naturalists also to consider the existence of previous biological provinces. Contemporary fossil data seemed to favour the idea of *no* provinces. Since most palaeontological research before this time had been conducted in Europe and the more accessible parts of North America, the emphasis on a uniform fauna and flora in the Tertiary period is perhaps understandable. Only naturalists who, for one reason or another, were against the doctrine of a central heat objected to this prevailing opinion. Such men were quick to advertise recent discoveries that did not fit in with the accepted scheme, and Robert

Jameson, in an effort to show that previous uniformity was improbable, took it upon himself to encourage, receive, and redistribute specimens from countries as yet relatively unexplored. Only as fossils from out-of-the-way places like Australia and South America trickled in to the museums of western Europe did it become possible for workers to think about provinces in a serious way and to suggest that perhaps there *had* been discrete associations in the past.

This growing awareness of previous biological provinces was given shape by the formulation of a so-called "law of the succession of types." The specimens that precipitated this rule were first reported in the editorial pages of the *Edinburgh New Philosophical Journal* for 1831. There, Jameson described two interesting discoveries, one a cache of fossil bones from caves in the Wellington Valley, near Sydney,[22] the other a remarkably complete and well-preserved skeleton of a *Megatherium* found by Woodbine Parish in the neighbourhood of Buenos Aires.[23]

The Australian bones—originally recorded by a military traveller, Colonel Lindsay, and then brought to Jameson's attention by Dr. Lang of Sydney—were received at Edinburgh and passed on to William Clift of the Royal College of Surgeons in London. Clift provided Jameson with a short account of their identities (they were extinct species of kangaroo and wombat), and a few reflections on the scientific implications of the discovery, which were written up in the last issue of the *Journal* for 1831. In this memoir "On the fossil bones found in the Bone Caves and Bone Breccia of New Holland," Jameson made the important statement that Australian fossils possessed the same structural distinctions as the fauna resident there today. The Australian biological province had clearly existed for longer than had been thought: "It follows . . . that New Holland was, at a former period, distinguished from the other parts of the world, by the same peculiarities in the organisation of its animals, which so strikingly characterise it at the present day."[24]

One year later the same fossils were examined by Joseph Pentland, also of the College of Surgeons. He confirmed that they were indeed specimens of giant kangaroos and other typically Australian species. "With a single exception," he wrote, "all the genera to which these fossils are referrable are now found inhabiting the Australasian continent."[25] This fact coincided with Pentland's belief that European fossils of the "Diluvian deposits" were special only to Europe and did not belong to a single worldwide population. In his view, both Europe and Australia had possessed a distinct and separate identity through geological time.

Woodbine Parish's *Megatherium* from Buenos Aires was examined by Clift in 1832 and described in the proceedings of the London Geological Society.[26] The anatomist E. W. Brayley was anxiously waiting for Clift's verdict. Only the year before he had asked whether this huge relative of the sloth family had always been confined, like its present-day representatives, to the New World.[27] Although Cuvier had asked the same question about an earlier, imperfect skeleton of the *Megatherium*, it seems that naturalists of the 1830s were particularly interested in this issue and that they saw themselves as being on the threshold of an important intellectual discovery.

Besides Pentland, Clift, and Brayley, further scholars contributed to the discussion. Charles Darwin and Richard Owen brought the subject to the very forefront of science by proposing a rule of the "succession of types," meaning that fossils normally belonged to families still inhabiting the general areas in which they were found. Darwin described this in 1838 in his *Journal of Researches*, pointing out that the structure of South American fossils had led him to consider the permanency of "type" through geological time.[28] It is doubtful, however, if Darwin did any more at first than recognise the obvious resemblances between living representatives of armadillo and his specimens of *Glyptodon*; the similarity of the carapaces of these two groups would be hard to ignore. Darwin's *Megatherium* was identified by Richard Owen in the first few months after his return from the *Beagle* voyage in 1836 and was reported to the Geological Society in 1837.[29] Until that time Darwin may well have thought he was the first to propound this rule: it seems probable that he only came across Brayley's account when searching for published descriptions of *Megatherium* specimens for his larger book, the *Journal*. However, Darwin put the facts down in such an articulate and effective manner that he should be given credit for publicising them more widely than any of his predecessors. The South American fauna, he asserted, was geographically distinct for long periods of geological time. Though this view was undoubtedly important for Darwin's developing ideas about evolution, it is worth noting that an apparent "succession" did not necessarily imply a transmutation of species. It could just as easily be interlaced with nonevolutionary hypotheses. When considered in its proper context the formulation of this law is seen as part of a debate over the recent history of the organic world, and not as the first step toward a transformation of biological thought.

To cap it all, some ten years later Richard Owen claimed that *he* had discovered the law, no doubt believing that it was his identification of Darwin's *Megatherium* and *Glyptodon* that generated the crucial evidence.[30] Owen, as is well known, did not believe in any sort of evolution, although he did advance his own scheme of an inbuilt progression through time that kept pace with geological changes. Like Jameson, he too was against any scheme that postulated one uniform fauna or flora for the Tertiary, although his reasons were more than just a dislike of the cooling earth theory.

Owen was well aware of the richness of the fossil record. As the self-styled "English Cuvier" he was at the very heart of British palaeontology, specialising in the anatomy of extinct quadrupeds, not the least of his achievements being the description of many recently excavated dinosaurs. It was virtually impossible for Owen to dismiss the evidence of nonuniform habitats in each geological period. In 1841 when he ushered in the word "Dinosaur" and elevated these "terrible lizards" to ordinal status, he clearly distinguished them from plesiosaurs and ichthyosaurs, marine saurians of the same era: dinosaurs, in his opinion, were basically land quadrupeds. These two great subdivisions of land- and sea-living lizards necessitated a diversified topography for the Mesozoic period.[31] Of course, Owen was not the first to think this: before him, experienced geologists such as William Buckland, Henry De la Beche, and Gideon Mantell had gone to great lengths to demonstrate what could be called the ecology of fossil populations (illustration 4). To Owen, however, it seemed that theories of the gradual diversification of the earth's surface were inextricably mixed up with Lamarckian ideas about progression. To combat the latter he attacked the former, asserting that the earth had always supported a spectrum of habitats and geographical regions.

In this Owen was supported by the other great palaeontologist of the day, Louis Agassiz. Like Owen, Agassiz opposed Lamarckian progression, a scheme which was then under renewed scrutiny because of Robert Grant's work in Edinburgh. Agassiz certainly believed in the residual heat of the earth and in a directional history of the organic world, but he would not admit that there was any gradual "evolution" or transmutation of forms from period to period. He too attacked the doctrine of one worldwide fauna, citing his own researches on fossil fish as evidence:

Illustration 4 Henry De la Beche's reconstruction (1831) of a Jurassic scene. 1 and 2 are Ichthyosaurs; 3 is a Plesiosaur; the remains that litter the sea floor indicated the original shells and bones that are today's fossil evidence. William Buckland's painstaking researches provided De la Beche with the fine detail of the scene. From F. T. Buckland, *Curiosities of Natural History*, 2nd ser. (London: R. Bentley, 1860), frontispiece.

> It would be, however, sometimes amusing, were it not actually distressing, to see the manner in which some geologists deal with fossils, considering them simply as the characteristics of certain rocks, and hardly yet dreaming that there may be such a thing as a special zoology of the different geological periods, and that during each, local faunae may have existed with peculiar animals etc.[32]

Agassiz was hardly tilting at windmills. In the 1840s the theory of a uniform population was still sanctioned by many eminent geologists, although the point in geological history at which this single fauna and flora was thought to have broken up was now pushed back in time to the upper formations of the Silurian system. The theory's continued vitality is best explained by the fact that much of contemporary stratigraphy depended on the appearance of "marker" species and on establishing an identity between fossils in disjunct beds. If identical forms could not be found, the palaeontologist searched instead for generic relationships and for the analogies between populations that went under the name "facies." These similarities—specific, generic, or merely overall likenesses—indicated that sets of organisms had lived under approximately the same conditions at the same time. The definition of a fauna or flora as groups of species existing under roughly the same conditions virtually demanded such an interpretation. Alcide d'Orbigny, the noted French palaeontologist, stated in 1843 that as the range of present-day species depends on a continuation of the same environment and temperature, so the absolute identity of fossils in separate lands leads to the conclusion that they shared a single climate. In a paper called "General considerations regarding the palaeontology of South America compared with that of Europe," he emphasised that every geological system in Europe preserves the same "palaeontological *facies*" as its counterpart in North America, although the similarities did not extend to the south. This general uniformity in each of d'Orbigny's twenty-seven systems depended on the uniformity of temperature. Only with the refrigeration of the globe and the erection of geographical barriers did d'Orbigny's fauna split up into local distributions. He believed this point occurred somewhere around the Cretaceous era.[33]

R. I. Murchison, while adhering to the same point of view, pushed the dividing line back to the upper Silurian. Although the universal nature of the Palaeozoic fauna was never really in question, Murchison conclusively showed that many of its species were spread very widely: several fossils of the upper Silurian in Sweden were not only identical with those of Great Britain but with those of North America as well. It was "a striking illustration of the wide diffusion of similar conditions in the early stages of the formation of the surface of the earth."[34] In support of this, his work on the Devonian system (more recent in geological time) demonstrated the increasingly diverse conditions under which Devonian animals lived.

Clearly, the second quarter of the nineteenth century was rich in palaeontological interpretation, and the geographical distribution of fossil species played a significant part in the continuing debate over the path the organised world had followed. Undoubtedly, this was because the subject had become associated with a complex body of mutually supporting theories, where the directional history of the globe suggested the idea of a sequential "advance" in the fossil record, coupled with a slowly diversifying topography and a gradually cooling earth. Geographical distribution supported many facets of this conglomeration of theories. For most geologists these interlocking hypotheses proclaimed an earth undergoing irreversible periodic changes—the "revolutions" first introduced by Cuvier. The subject of distribution therefore added depth to arguments based on a linear history of the earth.

LYELL, SEQUENCE, AND HISTORY

For this reason Charles Lyell (1797–1875) could hardly avoid examining geographical distribution in his masterly *Principles of Geology* (1830–1833). Unhappy with current explanations based on a directional earth history—although determined to retain a modified version of the central heat—Lyell postulated an endlessly fluctuating cycle of geological and organic change. In modern terminology he suggested a steady-state earth. As many scholars have noted, this view depended on a strong continuity between the present and the past. Lyell stressed that the processes of denudation, erosion, and volcanic activity had acted in the past at the same pace as today; that the geological and geographical transformations that could be seen in the stratigraphical table were still taking place, though at rates too slow for human eyes to see; that the members of the fossil record had been introduced and extinguished one by one, just as he supposed modern species to be created and exterminated. Having deliberately chosen a wide-ranging brief, Lyell left himself no option but to carry out a review of the whole spectrum of the earth and life sciences. Among other things he had to show that the history of animal and plant distribution was not a well-ordered series of changes matching progressive diversification of the earth's surface, steadily approaching the patterns (and the species) of today. For Lyell's purposes it had to be a history of never-ending fluxes, where each epoch supported a great number of distinct populations that expanded and contracted their boundaries as geographical barriers were removed or erected. The distribution of plants and animals had to be as complex in the past as it is today. And, according to the principles he was expounding, these topographical and chronological shifts were to be explained by causes common to both past and present, acting with the same unvarying intensity.

Lyell therefore turned to examine present-day patterns of distribution. Despite a great interest in the subject during the 1820s on the part of botanists like Augustin de Candolle, there were few comprehensive theories or rules of thumb which Lyell could follow. Those that did exist were used by geologists to further schemes of the earth of which he disapproved. So he probably saw tactical advantage in reformulating the study of past and present distributions into a coherent subject that tacitly reflected his own views. Certainly, in the second volume of the *Principles* he gave an excellent summary of the field before presenting concepts and methods that were almost entirely new to this area of thought and that proved to be of far-reaching significance. He brought to the study of animal and plant geography a deep sense of history—an appreciation of development over time, which was mirrored in the arrangement of today's organisms. He studied processes and thought about the dynamics of floras and faunas in conjunction with the continuing flow of geological change.

Lyell began his account of geographical distribution by recapitulating the handful of generalisations that had emerged from nearly one hundred years of research. Pouring ridicule on Linnaeus, he praised Buffon for discriminating between tropical species of the New and Old Worlds. He also admired the elder Candolle for identifying the large number of diverse stations that any one geographical region might present and commended Candolle's distinction between "stations" and "habitations"—a distinction Lyell understood as the difference between botanical topography and botanical geography. As with all his work, Lyell's survey tended to favour those people who shared his own ideas, or whose results could be interpreted in a suitably Lyellian fashion, at the expense of the rest. In this way he emphasised the importance of understanding the great diversity of the natural world. He went on to describe the various means by which dispersal takes place and the geographical barriers that effectively limit species to particular areas. Assuming that there were no barriers to dispersal and that species entered the fossil record as a single pair of individuals, Lyell believed that a particular organism could spread virtually all over the world, given enough time. By analogy, a whole flora or fauna could move across the globe if topographical and ecological barriers were removed. Lyell illustrated this by describing the anticipated movement of arctic plants into lower latitudes if the earth experienced a period of refrigeration. In the same way, alpine species residing at great heights on the mountains of Scotland and Europe would "immediately descend from these elevated stations to overrun the low grounds" if the climate of the polar regions was transferred to British latitudes.[35]

This thinking clearly linked geographical distribution into Lyell's larger scheme of climatic change. Anxious to refute the idea of a gradually cooling earth, he devised an ingenious theory which explained the past climates of the world without reference to any directional heat loss. He correlated climate with the disposition of continents and oceans, so that changes in the ratio of land and sea led to changes in the world's isotherms, currents, wind patterns, and the like, shifting seasons and causing long-term alterations in the annual distribution of heat. Particularly important was his suggestion that the elevation and submersion of large landmasses resulted in the conversion of equable climates into extreme ones, and vice versa, according to the quantity of land left above sea level. Lyell saw this as an endless sequence of changes, a cycle of climatic fluctuations romantically called the "great year," clearly based on a steady-state earth where huge sections of the earth's crust were elevated and depressed in a self-regulating and balanced system.[36]

For geographical distribution to fit this pattern it is plain that floras and faunas would have to be dynamic entities. Lyell therefore described a continuous fluctuation of living populations which expanded or contracted their boundaries as geological agents altered local topography and climates. No wonder, he remarked, that naturalists could find no regularity in geographical distribution when they thought of it purely in terms of the present-day configuration of land. On the contrary, there had been nothing but flux since the world began; a continuous interaction between the organic and the inorganic, where present-day arrangements are only a stage in the continuing flow of events.[37]

Lyell's writings on this subject were of great significance for the development of distribution studies. Their value lay not so much in what was actually said as in the concepts and assumptions that underpinned his approach to nature. These were absorbed and reiterated by Lyell's associates, Edward Forbes, Joseph Hooker, Wallace, and Darwin, each of whom made fundamental contributions to the field.

First of all—and Lyell was by no means alone in this—he took the important step of assuming that modern floras and faunas had all enjoyed lengthy histories. Lyell, in particular, understood today's associations as a result of past movements and volume II of his *Principles* was given over almost entirely to a thorough examination of the changes the organic world experienced at the hands of the inorganic. Other naturalists who gave existing floras and faunas the benefit of a past—among them Louis Agassiz, Alexandre and Adolphe Brongniart, and the two Candolles—did not lay as much weight on history as an agent of change as did Lyell. Even so, this view was seemingly confined to naturalists immersed in the geological sciences and in daily contact with theories relating to the probable *sequence* of events in the earth's past. It was here, in the study of geology and palaeontology, that naturalists familiarised themselves with methods based on conventional historical research: they looked for causes and effects, and for the forces that transformed one state of affairs into another over extended periods of time.

Such historical attitudes to natural phenomena were, however, a fairly recent innovation. Only since the Enlightenment had the earth taken on a "life" or existence of its own, independent of an interventionist Deity and far removed from the history of mankind and its Maker. Once allowed to be a separate body, no one could deny the earth had had a past and was about to experience a future. This sort of approach led natural philosophers to think of their existing surroundings as representing more than just the present: because the lands and seas were now thought to be constantly recombining and restructuring themselves, students of the earth realised that everything they saw was the product of time and process. Nothing was really new or pristine. "The past," declared William Whewell in 1840, "has been a series of events connected by historical causation, and the present is the last term of this series."[38]

Whewell's words were directly inspired by Lyell's treatment of geological history. The past, according to Lyell, was a string of continuous and consecutive changes, each geological event being part of a sequence of cause and effect. His steady-state earth demanded that more stress be laid on sequence than was previously the case in geological theory. Without recognising it for what it was, Lyell's precursors and contemporaries were dimly aware of this idea and had sometimes made use of it in attempting descriptions of the earth's past. Buffon, for instance, had pointed out that the progressive refrigeration of the globe had given the earth a sequential history, where each stage followed on from the next in a predetermined manner. And at a period rather closer to Lyell's, Cuvier had also drawn attention to the successive environments characterising the Paris basin as Tertiary seas periodically flooded the area.

Yet Lyell's idea of sequence was something more. Uniformitarian thought enforced an understanding of the gradual steplike nature of all geological change, and he postulated that all geological series represented artificial discontinuities in the history of the earth. There were, to him, no breaks in the geological table, and each stratum, formation, or system had gradually blended into the next by the continuing action of forces proceeding at the same intensity as those of the present day. He stated that everything to be seen in the geological record was in fact determined by what had gone before. Unlike Buffon, who had asserted that a relationship between epochs was necessary because the earth was experiencing a succession of states where A was followed in an orderly fashion by B and C, Lyell stipulated that every geological event (and hence every geological structure) was a rearrangement of the layers and resources then in existence. In the same way as today is yesterday's tomorrow, Lyell assumed that every occurrence in the history of the earth was at the same time both effect and cause: the effect of what preceded, and the cause of what was eventually to follow. In other words, every stage in geological history was part of an unbroken series or sequence of continual activity.

This is plainly an important view for biological theories. It was this notion, of course, that inspired Darwin to think of the gradual change or transmutation of one species into another over time. Yet it also stimulated Lyell's followers to set about integrating the twin studies of geology and distribution. Lyell wrote of the *motion* of the earth's crust and of its organisms. He saw the distribution of living beings as a succession of extensions and withdrawals according to structural changes of the earth, as a series of dislocations and reconnections of ranges, and as an unending mixture and isolation of organisms. His descriptions of living populations were based on an interplay between organic and inorganic forces. Furthermore, by its very gradualism, Lyell's account gave animals and plants time to disperse or retreat in the face of favourable or unfavourable conditions at an entirely natural pace. Certainly there was room in his system for major geological upheavals (he would not, for instance, have denied that the sudden appearance of Monte Nuevo was an event of dramatic proportions), so long as these events did not exhibit any intensity of action not found in the present. But Darwin and Forbes, for example, took him to mean that the elevation and subsidence of continents was usually slow enough for living beings to act in concert with these changes. This last view encouraged both Darwin and Forbes to postulate that floras and faunas had spread naturally over wide areas which may not now be connected together. From such suggestions the roots of modern biogeography were to emerge.

STATISTS AND HISTORIANS

A comparison of the investigations of geologists with those of botanical or zoological geographers at this time gives rise to a number of interesting distinctions. Geographical distribution seems to have meant different things to different groups of naturalists.

Depending on whether they came from a "geological" or a "taxonomic" background, the way in which nineteenth-century naturalists approached geographical distribution varied in a characteristic manner. Each discipline gave its practitioners a certain way of looking at things, a certain way of thinking about the evidence, and a distinctive set of assumptions. These predilections guided naturalists when faced with a relatively new field of scientific inquiry. In consequence, the style of research and the kind of hypotheses proposed echoed the past experiences of each worker.

Broadly speaking, geographical distribution was divided into a study of processes on the one hand and of patterns on the other. An emphasis on processes was clearly seen in the work of the geologists and other naturalists discussed in the present chapter. They investigated the development of regional associations over time and the qualitative changes found in successive floras and faunas. Under the circumstances they could do little else. Current hypotheses about the evolution of the physical features of the earth and the linear path of history provided these geologically minded scholars with potent models for the interpretation of the fossil record. As a result, the fossil record was explained in terms of a combined geologico-organic advancement through time. In turn, this reinforced the notion of life being analogous to a series. With hindsight it seems almost inevitable that naturalists did what they did. Facts from present-day distribution were expected to support explanations concerned with the *history* of the earth; the data was therefore ordered into a consecutive history of its own. From Buffon to Agassiz, geologically minded naturalists were preoccupied with the idea of a chronological succession of phenomena and their causes. This concept of a series of changes was transformed into a continuous story by Lyell's authoritative work. History had been introduced to the study of animal and plant distribution.

How different to the work of natural historians, where systematists such as H. C. Watson investigated the topographic arrangement of species and the various proportions that different classes held to each other inside well-defined boundaries. With the aid of elementary statistics, botanical geographers endeavoured to describe the internal structure of assemblages and to compare one set of species with another. Their thinking was dominated by the idea of *groups* of living beings: the existence of groups, their composition, characteristic species, boundaries, and physical requirements. In this sense natural historians were "pattern" men, concerned with tracing out the multiple arrangements of existing faunas and floras. They could be called the topographers of the natural world.

In effect, natural historians examined geographical distribution as it occurred in topographical space, while geologists concerned themselves with distribution over time. But the distinction here is more than a simple dichotomy between the two. It seems that these twin approaches to natural phenomena influenced the character of the scientific proposals that were being made.

Topographic natural historians, of whom Watson is the prime example, chose to describe biological provinces and groups of living beings in demographic terms. They itemised the "statistics" of each population as if conducting a census. To some extent this was a legacy from the past, in that there was a long tradition of including biological facts in sociopolitical surveys of separate nations. The term "statistics," as previously mentioned, originated in a comparative description of states. Hence there is nothing unusual in finding an analogy being drawn between the distribution of animals and plants and that of human beings. Yet the nineteenth-century explosion of interest in the size of the British populace, and in numbers generally, contributed far more to the establishment of a "topographical" science than this simple analogy. Censuses were commonly regarded as the most efficient and desirable way of collecting facts; and facts were what most scientists of this reforming age wanted to have. In the belief that a field of inquiry ought to be constructed on Baconian principles, botanists and zoologists of the second quarter of the century restricted their activities to the gathering of highly detailed information. The data was subsequently arranged—according to the model of a census—in tables ostensibly free of any explanatory hypotheses. This was what "botanical statists," as Watson called himself, were supposed to do.

On the other hand, geologists were expected to research into the past. By 1830 or so theoreticians were freely reconstructing the history of the earth. They self-consciously organised their findings into a historical—or developmental—sequence. "Geologists are the historians of the natural world," emphasised Sedgwick in 1830.[39] His words drew a sharp line between the activities of geologists (who interpreted the "testimony of the rocks") and those of natural historians, who merely described the details of the present. The one was creative where the other was simply an onlooker. James Cowles Prichard made a similar distinction between the historical and descriptive sides of anthropology, pointing out that natural history, by definition, was a purely factual affair whereas his reformulated discipline of ethnology was a science of the history and development of human races.[40]

Clearly there were ulterior motives in divorcing the historical sciences from the descriptive ones: to do so effectively disassociated Whewell's "palaetiological" disciplines (good) from their less sophisticated past (bad), and emphasised their modernity.[41] Nevertheless, this rough characterisation of two separate "styles" of scientific work is accurate enough. Natural historians of the early part of the nineteenth century did describe the appearance and classificatory relations of existing organisms, while geologists confined their studies to developments over time.

Making allowances for personal differences, this same dichotomy can be seen in the field of geographical distribution. It is hard to say at what point individual tendencies can be called a "style," especially when early-nineteenth-century figures show little desire to group themselves into "schools" of any kind. Equally, the division between "taxonomists" or "natural historians" and "geologists" is frequently blurred by the comprehensive interests of men such as Darwin, Hooker, and Lyell. Yet it is well known that social experience influences the scientific work of individuals. The men discussed in this chapter shared a common experience of the geological sciences, and their work in geographical distribution seems to reflect this. Similarly, the naturalists considered in an earlier chapter shared an interest in the description of the existing world and in the relations between its various parts. One group came up with a "historical" interpretation of modern geographical distribution; the other with a "topographical" or "statistical" mode of investigation. But the dynamic and historical approach introduced by Charles Lyell proved the more enduring. Promoted by his disciples, the Lyellian mode of thought was applied most dramatically—and most effectively—to the ever-present problem of the creation of life.

FIVE

The Earth as a Chessboard

Long before the 1840s, the Noachian model of species spreading out from one common location had been replaced by the idea that animals and plants originated in the general area in which they are now found. The latter view was formalised in 1820 with Augustin de Candolle's definition of botanical "regions" as areas where species are endemic. Indeed, it had been stated even earlier in Buffon's rule that animals of the tropics were native to either the New or the Old World.[1] By the time Lyell wrote the *Principles of Geology* these regions—areas characterised by endemic forms—were generally called "centres of creation," a term roughly synonymous with the modern expression "biological province." The inhabitants of a centre of creation were seen to be distinct from those of other associations and especially suited for the area in which they appeared. They were thought to have been created there.

Within this paradigm there was room for several hypotheses about the first appearance of living beings. Were the original representatives of a species a single reproductive couple or an entire breeding population? Had whole groups of species appeared at the same time or did one form follow after another in sequence? Were the species of a province created in one spot—a kind of centre or focus of creative activity—or dotted about throughout the region? All these suggestions were aired during the first half of the nineteenth century. Robert Brown, for example, considered that there was a "focus of creation" in eastern Australia and another in the west. The two halves of the Australian flora were thought to have appeared simultaneously in these restricted foci, each species being represented by a single reproductive pair. From there they increased in number and dispersed over the surrounding countryside. After Brown's time the tendency was to multiply the number of sites where creation may have taken place. Alphonse de Candolle suggested that there could not have been a single focus of creative activity in a province since aboriginal pairs were more likely to have sprung up in different places within the general boundaries of the region. He was uncertain if this happened all at once or over an extended period of time, as Charles Lyell had proposed. Lyell's species appeared in separate places, coming into existence one by one, in an irregular, piecemeal fashion. Subsequently, this proliferation of sites within biological provinces was extended by Louis Agassiz, who believed that all species of a region were created at the same time, each and every one distributed over the full extent of their present ranges. He had abandoned the concept of "pairs" in favour of many individuals of a species springing into existence simultaneously.

But there was little concrete evidence on which to base a choice between theories. Naturalists hardly expected to witness the actual birth of a species, though they did expect the phenomena of geographical distribution to reveal important clues as to how it happened. Yet the clues that had emerged perplexed scholars by pointing in all directions at once; they could be interpreted in any number of ways. And an already difficult situation was made worse by repeated discoveries of the same species in two or more biological "provinces." Robert Brown thus gave a list of 150 European plants that could also be found in Australia; had they been created twice, once in Europe and again in the antipodes? The definition of a centre of creation seemed to require this interpretation. All the other species living in Australia were assumed to have originated there, so why not the so-called European forms as well?

Of course many of these problem species could legitimately be explained as cosmopolitan or otherwise widely ranging organisms. Some, such as insular representatives of species found on neighbouring continents, had obviously migrated across the intervening straits. But others were apparently inexplicable. They appeared in floras and faunas very far apart, with no further biological features in common. Augustin de Candolle carefully distinguished such perplexing plants from more normal species, calling them "sporadique." All other species, he pointed out, were endemic by nature, that is to say, confined to only one region of the globe. Why sporadic species should be different he did not say, arguing that the problem could hardly be solved until the origin of ordinary plants had been sorted out.[2]

To explain sporadic species—or disjunct species, as they are now called—the only possible answer seemed to be a hypothesis of multiple creation. An organism that appeared in two otherwise distinct communities must have been created twice. When all natural explanations are exhausted, advised the entomologist Lacordaire, the scientist is forced to accept this conclusion: "in such a case we are almost obliged to admit that this separation has taken place with the origin of that species; in other words, that it must have had a multiplied origin."[3]

Multiple creation, in itself, was not such an unreasonable suggestion. If naturalists were prepared to believe in the supernatural creation of species, then it was as logical to believe in double creations as to believe in single ones. But this invited interpretations of the living world that ran against all tradition, against the mainstream of scientific thought, and, in some cases, against religious belief. To accept dual origins for certain species meant a number of things. It meant that these species, at least, did not originate in pairs of individuals, a paradox that led Agassiz to the proposal that *all* species must have begun as full breeding populations. It meant that species that were created in two different places may also have been created at two different times, making nonsense of the idea of extinction and overturning the greater part of contemporary geological theory. And although the vast majority of naturalists believed that God was the author of *all* species, whether they were created singly or in masses, multiple origins could nevertheless provide theoretical backing for some to interpret the living world as the mere product of its physical environment. The botanist Sprengel dispensed with a Creator and advocated this last view, arguing that species "common to the most distant countries" led to an inescapable conclusion: that "the vegetable world has neither descended from one common birthplace, nor diffused itself from one country into another; but that vegetation is in every case the product of the joint influence of temperature, soil, and the particular composition of the moisture of the earth."[4]

Understanding the origin of what are now called disjunct species was therefore crucial to the larger question of the first appearance of species. The doctrine of multiple creation threatened to overthrow all the recent advances made by geologists and natural historians in recording and translating the fossil record, in the comparative anatomy of past and present organisms, and in the study of geographical distribution patterns. To defend these professional achievements it was necessary to demonstrate that multiple creation was not true; that sporadic species could indeed have originated as pairs of individuals, introduced at one time, and in one place. It was with these deeper problems in mind that Edward Forbes, Charles Darwin, and Joseph Hooker approached the field of geographical distribution.

FORBES AND SINGLE CREATION

Now barely remembered by historians, the Manxman Edward Forbes (1815–1854) was in his time a figure of considerable stature. In geological circles he stood on a par with Charles Lyell, Roderick Murchison, and Leonard Horner, all of whom were close professional friends. Thomas Henry Huxley, another friend, thought Forbes had a "great intellect"; he was an "acute and subtle thinker" who understood many diverse fields of inquiry and commanded respect and affection.[5] He was, in short, a force to be reckoned with.

Besides making his name known to British geologists through a series of papers on conchology delivered to the newly established British Association for the Advancement of Science, Forbes had been taken up by Roderick Murchison, one of the founding fathers of the Association. It was through Murchison's good offices that Forbes was enticed to London from Edinburgh, where he was half-heartedly studying medicine. The Edinburgh days were instead spent absorbing the latest in post-Romantic German philosophy and developing his own transcendental interpretations of nature. He had at that time expected to inherit enough money to pursue his interests as a gentleman-naturalist, but when his hopes failed to materialise he turned instead to science for his livelihood. By 1841 he had obtained a congenial if temporary post as naturalist to Captain Thomas Graves. They sailed to the Aegean, where he indulged a penchant for marine dredging, watercolour sketching, and archaeological exploration among the ruins of Lycia.

Unfortunately, the surveying trip was cut short when Forbes contracted malaria. In any case his London friends had by then found him a (poorly) paid position lecturing in botany at King's College. To supplement his income he managed to get a second appointment as curator to the Geological Society's collections, again through friends, of whom perhaps Lyell and Horner were most interested in his career.

During the next ten years Forbes went from strength to strength. In 1843 he presented the results of the Aegean survey in a paper on the "Mollusca and Radiata of the Aegean Sea," a title which belied the scope and intellectual richness of the work. It was here that his first major statements about distribution were articulated. On the basis of personal research, Forbes asserted that the depth to which a marine plant or animal could range was proportional to its extension in a geographical direction: "The extent of the range of a species in depth is correspondent with its geographical distribution."[6] The mollusc *Arca lactea* was a true native of faunas at all depths in the Aegean and also stretched north to British waters and south beyond Gibraltar. Its great range implied some capacity to withstand a wider variety of conditions than most species. Second, in the same paper, Forbes stated that there were zones on the sea floor similar to the zones found on a mountainside, each region containing a flora and fauna analogous to its latitudinal parallel. "Parallels in latitude," he explained, "are equivalent to regions in depth, correspondent to that law in terrestrial distribution which holds that parallels in latitude are representative of regions of elevation."[7] Both of these axioms were to prove important for his later work on disjunct species.

In the early months of 1844 Forbes gave a Friday evening discourse at the Royal Institution, simplifying and expanding the Aegean paper into "The light thrown on geology by submarine researches," which was praised by Horner in his presidential address to the Geological Society the following year.[8] The paper also ensured his appointment in 1844 to the newly formed Geological Survey of Great Britain under the direction of Henry De la Beche. From this vantage point Forbes published the long memoir for which he was remembered by his contemporaries: "On the connexion between the distribution of the existing fauna and flora of the British Isles, and the geological changes which have affected their area, especially during the epoch of the northern drift" (1846).[9] It was the most detailed attempt so far to correlate the present-day patterns of floras and faunas over Europe with geological upheavals and submersions of the past.

After this contribution honours were heaped upon Forbes as he continued to write, lecture, and collate fossil collections; a rise to fame which culminated in his presidency of the Geological Society in 1853 and, in the following year, his election to the chair of natural history in Edinburgh. He succeeded his former teacher, Robert Jameson. For both these posts he was backed by the most eminent naturalists of the day. The move from London to Edinburgh was mourned in geological circles, but his friends were even more upset when they learnt of

Forbes's death—from suspected kidney failure—only a few months later.

Forbes intended to explain the origins of British plants and animals in his 1846 memoir "On the connexion, etc.," noted above. The British flora, in particular, formed the central feature of an argument for the existence of what he called "specific centres" or, to paraphrase, the creation of species in aboriginal pairs at one particular location. He believed that only two individuals for each species had first appeared; from the single point that they may be assumed to have occupied, they had extended their range more or less concentrically as numbers increased through normal reproduction. There were several reasons for believing in this process, not all of which were made clear in the memoir. First, of course, it indirectly reflected the biblical story—a potent argument for a religious man like Forbes. Furthermore, it had the elegance and simplicity characteristic of the better kind of scientific hypothesis. However, there was empirical evidence as well. Pairs of modern individuals introduced into new habitats colonised areas in just such a way; in a properly actualistic theory there was no reason to suggest that things had happened any differently in the past. The same phenomenon could be seen on a larger scale in the natural arrangement of floras and faunas, where groups of species extended their ranges after the demolition or bridging of geographical barriers. A capacity for expansion signified that populations were primarily restricted by mechanical devices, another argument for the radiation of individual species from an original "creative" spot. And finally, for Forbes there was an attractive symmetry in supposing that species were "born" in the reverse manner to their death. He was among the first to propose that species en route to extinction gradually became less numerous in individuals and more restricted in range until only two individuals finally remained.[10] If these failed to reproduce then the species became extinct. The argument was not so much a corrective to the theory of catastrophism—the total annihilation of many individuals all at once—as a convoluted plea for the idea of single creations and extinctions.

So Forbes's object in the paper "On the connexion, etc." was plain. He wanted to show that British organisms could best be explained on the supposition that each species began in one spot and as a pair of individuals. He observed that the British population consisted of several floristic or faunal "elements," all of which originated in neighbouring countries. (Forbes was one of the earliest scholars to divide a population into its component elements in this fashion, although he did not use that particular expression. The term "phytogeographical element" was not formally introduced until 1867, when the botanists Areschong and Christ proposed it simultaneously.)[11] The species that made up these elements arrived in the British Isles at different times and from different directions, according to the fluctuations of European land and sea. Geological movements explained why, for instance, a Dutch species predominant in East Anglia did not feature in the plant life of Wales; the diffusion of the Dutch flora westwards had been halted before any of its frontline members reached that area. Again, Ireland was cut off from mainland Britain at an early stage, thereby receiving only a handful of colonists from the floras and faunas spreading over the rest of the country.

There were many facets to Forbes's account. Geological, zoological, and botanical information was freely interchanged, building up a complex and comprehensive history of British species. It was all based on the thesis that the appearance of species depended on their "transmission, through migration, over continuous or closely contiguous land." Forbes's most controversial point was his proposal for a new "Atlantis" in the Miocene period, linking the Iberian coast with the Azores (roughly 1,000 miles distant) and thence with Ireland (another 1,000 miles). The landmass was supposed to account for the existence of little more than a dozen Asturian plants, belonging mostly to the Saxifrage family, in the southwest of Ireland. But such an enormous and recent upheaval to explain only twelve species generated much righteous criticism from more economically minded geologists. "I think it a bold step," complained Darwin, "to sink into the depths of the ocean, within the period of existing species, so large a tract of surface. . . . I must say I should like better evidence than the identity of a few plants, which *possibly* (I do not say probably) might have been otherwise transported."[12]

ARCTIC AND ALPINE FLORAS

Criticisms apart, Forbes was congratulated on other aspects of the same argument. The best and most striking point in his defence of single creation was that of arctic plants—mainly Scandinavian species—which appeared on British uplands and the summits of continental European mountains.

These peculiar distribution patterns were widely known by the middle years of the nineteenth century. Arctic regions had been botanically explored by Linneaus, Gmelin, and Pallas, among others, and were now under renewed scrutiny from travellers in search of additional trade routes. The mountain ranges of western Europe had also been scientifically explored, to a greater or lesser extent, since Conrad Gesner first climbed Mount Pilatus near Berne in 1555. Alpine plants showed the same structural adaptations to their physical surroundings as those of the arctic circle, having a stunted growth pattern, a general thickening of the green parts, and a tendency to form cushions, tussocks, and rosettes. The superficial analogy to arctic plants was widely noticed and attributed to the severity of local conditions or the poverty of physical resources. Scholars were not slow to recognise a parallel between zones of elevation and latitude: first recorded by Tournefort in 1685–1689, the analogy was used by Linnaeus in a description of Mount Ararat and by his contemporary Albrecht von Haller in a *Vorrede zur Geschichte Sweizerischer Pflanzen* of 1772, before being codified by Humboldt in the beginning of the nineteenth century. Each high mountain of the tropics was thought to carry the vegetation of a single hemisphere on its slopes, culminating in an arcticlike flora at the summit. (Yet this idea did not spread much beyond academic circles: even as late as 1849, when Mount Kenya was discovered, the public found it hard to believe that permanent snow existed on the equator.)

With increasing sophistication of taxonomic methods, longer and more adventurous field excursions, and diversifying printed resources, European botanists found the parallel between elevation and latitude even more plausible than they had assumed. Many of the plants were genuinely common to both polar and montane regions. Of the 600 or so species found near the snow line in the Swiss Alps, no less than 270 reappeared in the extreme north, among them several species of *Silene*, *Saxifrage*, and *Salix*. The distribution of the animal kingdom echoed these features: Latreille noted that the insect fauna of montane Switzerland was dominated by Swedish species, singling out the beetle *Prionus depsarius* for special mention.[13] Nor was the connection between only Switzerland and Sweden. The instigation of field trips in the universities of Glasgow and Edinburgh in 1821 by their respective professors of botany, W. J. Hooker and Robert Graham, led to the first detailed examination of the Scottish Highland flora and the discovery that Scottish mountains supported a great number of Scandinavian plants. Accompanying Graham on many of his excursions was his pupil Hewett Cottrell Watson. Although not every subarctic species was present in the British Isles, as Watson pointed out, nearly every Highland plant could be found in the Scandinavian peninsula. These hardy forms were also present in decreasing numbers along the mountain ridges of northern England and Wales, although here they were more likely to be accompanied by modified lowland species towards the south.[14]

The physical conditions of these upland sites were so similar that, under other circumstances, the plants would have been considered members of one continuous flora. It seemed as if the boreal flora was successively diluted as the distance from the arctic circle increased. Yet mountain summits were clearly separated from each other by lowlands in which no montane plant could be expected to live, and over which no naturalist would seriously suggest they had migrated. Moreover, the Alps and Sweden were divorced by some two thousand miles of even more inhospitable terrain for the transfer of species. How had they got where they were? It was very easy to assume with Gmelin or Sprengel that they had been created for—or spontaneously developed from—each particular site. It was easy to assume that multiple creation had taken place. This was the view Edward Forbes set out to repudiate.

Forbes proposed that the identity between the floras of Scandinavia, Scotland, and Switzerland was a direct consequence of the subsidence and subsequent elevation of the whole of northern Europe during the period he called "glacial," or that of the "deposition of the northern drift."

Following Lyell, and in accordance with a great deal of apparently confirmatory evidence, Forbes believed that much of the northern part of the earth had experienced very cold conditions and an extensive subsidence of land in the recent geological past. During this period glaciers were thought to occupy the high ground while snow and ice more or less covered the remaining lowlands; oceans etched out submerged river valleys, rounded off the sharp features of drowned continents, and swept along in their currents great tides of silt, gravel, and clays, these eventually being deposited in irregular and jumbled beds known to geologists as "diluvium" or "drift." At the same time, icebergs floated at random in the "glacial" sea, scratching rocks and depositing boulders at irregular intervals. This was Lyell's version of

British "Drift Theory,"[15] an elaborate alternative to the "Ice Age" proposed by Louis Agassiz in 1840. The latter hypothesis required a universal ice-sheet, formed by the extension of glaciers far beyond their existing limits. On their retreat, glaciers were supposed to have gouged out or deposited all the features now associated with the Pleistocene age.[16] Agassiz's theory was poorly received in Britain—although his ideas about glacial erosion were welcomed—the main reason being that geologists already possessed an apparently viable theory based on water.[17] Though Forbes used the term "Pleistocene" for his cool and icy period, he did not at any time follow the *Eiszeit* part of Agassiz's proposals. Not the least reason for rejecting this ice-sheet was Agassiz's contention that the Alps had been thrown up at the end of the cold period; it was essential for Forbes's theories of distribution to interpret them as older mountains, predating the cooler temperatures.

The lack of any concept of an Ice Age or ice-sheet did not materially retard Forbes's interpretation of Pleistocene phenomena. However, it did encourage him to concentrate on the watery parts of Lyell's scheme. The existence of an extensive but shallow sea over much of Europe, created by the hypothetical subsidence of the continent by a few thousand feet, was crucial to his explanation of organic distribution at that time. He suggested that the mountains of Scotland, parts of the Pennine Chain and of Wales, all known to be considerably older than the "glacial" period, were for most of that epoch nothing more than low-lying islands surrounded by an icy sea. At the same time the land was colonised by Scandinavian plants (and the sea by Scandinavian molluscs), each species expanding its geographical range as the environment became suitably cool and severe. If, as Forbes supposed, the great sea stretched as far as the Alps and central Asia, Scandinavian species could have dispersed over these regions as well—although in fewer numbers because of the great distance and relatively short period of time. Then, with the subsequent uplift of the seabed at the end of the Pleistocene and the alleviation of harsh and cold climates, Forbes suggested that the mountains were converted back to their original height and form, lifting the arctic plants up to two or three thousand feet in Britain, more in Switzerland. Only plants that managed to remain at or reach a height that presented suitably cool conditions would survive this elevation. Once there, there they would stay, surrounded by a "sea" of lowland species. In similar fashion, boreal species of shells were left stranded at the bottom of deep-sea trenches while the rest of the seabed was elevated; these too formed outliers of arctic species surrounded by the molluscs of temperate climates. Forbes explained:

> Now it was during this [glacial] epoch . . . that Scotland and Wales, and part of Ireland, then groups of islands in this ice-bound sea, received their alpine flora and a small portion of their fauna. Plants of sub-arctic character would flourish to the water's edge, but when a new state of things commenced, when the bed of the glacial sea was upheaved, its islands converted into mountains, its climate changed, and a suitable population of animals and vegetables diffused over its area, the plants of the colder epoch survived only on the mountain regions which had been so elevated as to retain climatal conditions similar to those which existed when those regions were low ridges or islands in the glacial sea.[18]

Here was an elegant explanation which introduced nothing that naturalists did not already know or take for a fact, and yet one which synthesised accepted concepts into something significantly new. Forbes took Lyell's period of the "northern drift" and the associated phenomena and postulated the environmental conditions that must have accompanied these changes; he took Humboldt's rule of the zonation of mountains, where the summits were in many respects physically analogous to true arctic regions; and he married them to the idea that floras and faunas spread while the earth's surface moves up and down. The result was a simple but effective hypothesis.

With his account of arctic-alpine species Forbes demonstrated that unremarkable processes of geological activity, in conjunction with the natural extension of floras and faunas, could explain the origin of disjunct species. Widespread species merely suffered an interruption of their range and were left as a series of isolated populations. With a mechanical explanation like this there was no need for hypotheses of multiple creation.

Geologists such as Roderick Murchison, Leonard Horner, and Charles Lyell were uniformly delighted with Forbes's elegant combination of geological dynamics and biological movement and envisaged the birth of a new science of geographical distribution based on the idea of motion. Murchison praised the theory of "the origin of Alpine Floras distributed far apart" at the British Association meeting of 1846 at Southampton and claimed these views of Forbes's "will introduce a new class of inquiries into natural history, which will link it on more closely than ever to geology and geography."[19] Horner lectured the

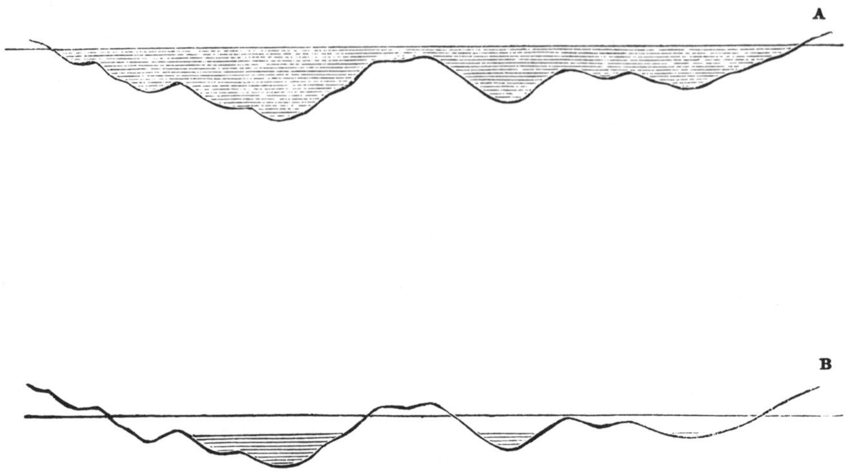

Figure 8 Edward Forbes's diagram of the changes of level affecting the distribution of marine animals. (A) represented the British sea floor during the "glacial" epoch, colonised by boreal and arctic faunas. (B) showed areas where arctic animals were stranded at the bottom of deep-sea trenches after the elevation of the sea bed.

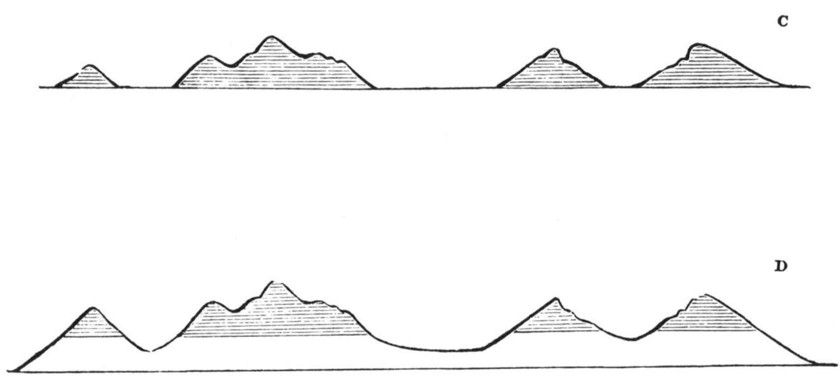

Figure 9 Forbes's companion piece to figure 8, showing the changes of level undergone by arctic plants. (C) represented a group of islands, all that remained of Britain during the "glacial" epoch, colonised by northern and Scandinavian plants. (D) Subsequent changes of level converted these islands into mountain summits, on which the arctic flora was preserved and isolated. Temperate plants occupied the foothills.

From E. Forbes, "On the connexion between the distribution of the existing fauna and flora of the British Isles, and the geological changes which have affected their area, especially during the epoch of the northern drift," *Memoirs of the Geological Survey of Great Britain* 1 (1846): 388–89.

Geological Society on the merits of this "admirable example" of geological reasoning in what he believed to be an "interesting and attractive" memoir, full of "suggestive" and "great views."[20] And Lyell was so gratified by Forbes's extended usage of his theories of local climate change, geological elevation, and the topographical movements of living beings that he appears to have borne no grudge that Forbes had failed to give him the proper credit that was his due.[21]

DARWIN FORESTALLED

Charles Darwin, however, was disappointed at being "forestalled" by Forbes.[22] Writing privately in a sketch expounding natural selection in 1842, and at greater length in 1844,[23] he had proposed an almost identical theory to account for alpine floras. Later, these ideas were written up for his "big species book" (1856–1858)—the manuscript was never published in his lifetime[24]—and eventually printed in the *Origin of Species* (1859), with the acknowledgement that Forbes had had the idea first.[25] There was thus no controversy over priority nor any spirited defence of minutiae of the kind that litter the history of science. The similarities were entirely due to the two men's enthusiasm for Lyellian geology and for dynamic interpretations of the structure of the earth correlated with the distribution of its organisms. The double formulation owed much to the geological ideas with which Lyell furnished Forbes and Darwin at the start of their respective careers.

Despite a commitment to evolution, Darwin's views on the history of floras and faunas were not all that different to those of Forbes. Instead of requiring species to be *created* as pairs of individuals, Darwin proposed that they *evolved* from parent species in ones and twos. Darwin's species originated in a particular location, increased in number, and extended their range in exactly the same way as Forbes's species did; Darwin's floras and faunas, just like Forbes's, ebbed and flowed over contiguous regions. Of course, Darwin's communities were liable to change, but to all intents and purposes they could be discussed as one unit over time.

To refute suggestions of multiple creation it was necessary for Darwin, as for Forbes, to explain disjunct species as the unexceptional result of natural causes. It was not so much the "creation" part of the proposition that annoyed him as the implications of the word "multiple." Like Forbes, Darwin believed the evidence of geographical distribution to be overwhelmingly in favour of species beginning as a single pair. If this was not true then the whole of natural selection theory was at risk. There could be no natural selection if large numbers of individuals changed into a new species in concert, for the essence of Darwin's idea was that selective forces worked on the individual differences among organisms, that novelties emerged one by one, gradually permeating the general population according to their adaptive success. Darwin had to believe in pairs (or in very small numbers) as the source of new organic types. Up to a point, therefore, arguments for single creation substantiated his own case for evolutionary theory. Where he differed from Forbes, Lyell, or Joseph Hooker was in his rejection of the assumption that an origin in a single pair necessarily implied a permanence of type. Hooker, for example, carefully set out in 1853 all the reasons for believing in the constancy of species over time,[26] noting that the idea of single creation must be connected with a certain permanency of form. Otherwise how could the same species still be in existence at the present day? Darwin alone seems to have been able to separate the two notions, and thence to reject the permanence of type.

In the first sketch of his theories of natural selection and evolution, drafted in 1842 and consisting of little more than a series of *aide-mémoires*, Darwin introduced the arctic-alpine problem into a general discussion of distribution to show that changes in climate or altitude could affect the ranges of species. He needed to establish that there were sufficient alterations in the physical configuration of land to trigger off all the rearrangements of species that might perhaps "unsettle" the "constitution" of species and make them vary. Here, he was concerned with fluctuations of topographical level that would induce some organisms to climb up or down mountains in order to remain in suitable climatic zones. Obviously basing his account on Humboldt's dictum of the zonation of mountains, Darwin envisaged climatic belts moving higher or lower, up or down the sides of hills, as the land itself subsided or was elevated by geological activity. The plants of each climatic belt would necessarily follow suit. If the land subsided to any great extent, the unfortunate species at the top of a mountain would clearly not survive, for there was nowhere higher for them to go in search of cool conditions. The social relations of the species of that area would be permanently affected. Building again on Humboldt's parallel between elevation and latitude, Darwin suggested in the later parts of his sketch that shifts in mountain floras would be paralleled by similar changes of climate over a continent. The rearrangement and extinction of plants on a mountain undergoing changes of level would be found also on a continent experiencing comparable changes in its climate.[27]

All these points and more were to be read in volume II of Lyell's *Principles of Geology*. In the middle of a long section describing the many changes that the earth's surface had experienced, Lyell stated that alterations in the level of mountainous lands would result in the movement of plants up and down their slopes and that analogous changes of climate (over time) on the "flat" would lead to the extension or retreat of arctic floras, or indeed of any flora, into different latitudes.[28] His purpose was simply to demonstrate that extinction was part of the regular order of things; but his words had a more insidious effect in sparking off Darwin's interest in disjunct populations.

In his *Essay* of 1844, Darwin proposed that, during the period of snow and ice at that time acknowledged as a geological certainty, cool European climates enabled arctic plants to move southward from their usual haunts to colonise whatever land was available. With a later change in temperature for the better, these species would be forced to retreat either back to the arctic circle or upward, toward the mountaintops, in pursuit of cooler conditions. The organisms remaining on elevated lands would be, in effect, little islands of arctic species marooned in a sea of temperate and lowland plants. Darwin noted that:

> Now we may with confidence affirm, from the number of the then floating icebergs and low descent of the glaciers, that within a period so near [to the present] that species of shells have remained the same, the whole of central Europe and of North America (and perhaps of Eastern Asia) possessed a very cold climate; and therefore it is probable that the floras of these districts were the same as the present arctic one. . . . At this period the mountains must have been covered with ice of which we have evidence in the surfaces polished and scored by glaciers. What then would be the natural and almost inevitable effects of the gradual change into the present more temperate climate? The ice and snow would disappear from the mountains, and as the new plants from the more temperate regions of the south migrated northward, replacing the arctic plants, these latter would crawl up the now uncovered mountains, and likewise be driven northward to the present arctic shores.[29]

So, taking Humboldt's analogy between height and latitude, Darwin reasoned that arctic plants were just as likely to move *up* as they were to return north under the pressure of changing temperatures. And he took Lyell's statement of climatic changes accompanying changes in vegetation to suggest that when Europe experienced very cold conditions it supported a flora suited to the snow and ice. Like Forbes, he assumed there was a sort of double flux—a movement of vegetation and climate simultaneous with the activity of geological processes.

But, and this is a notable point, Darwin did not suggest that Europe was necessarily subject to elevation and subsidence of land relative to sea level. His account did not mention the hydrographical configuration on which Forbes had insisted—that European mountains projected out of a shallow "glacial" sea and were the only available land for plants to inhabit. Nor did Darwin write of plants "flourishing to the water's edge" as did Forbes. For the major difference between Forbes's and Darwin's proposals was that Darwin had no need of the sea or of the watery configuration of drowned continents that contributed to the isolation of Forbes's arctic plants. There does not seem to be any particular reason for this divergence of opinion; it can hardly be attributed to any changes in the status of Lyell's drift theory, since geological opinion was still at this time firmly for water and against ice-sheets. Indeed, Darwin was a leading proponent of water-borne ice, an essential feature of Lyell's hypothesis. Perhaps Forbes's interest in the distribution of marine molluscs encouraged him to think about oceans and their geographical position, in contrast to Darwin's preoccupation with plant cover. Or indeed Forbes may have interpreted Lyell's map at the end of the *Principles* (showing regions that had been under water at some time or another during the Tertiary period) too literally, and so envisaged all areas as submerged simultaneously.

More significant from the historian's point of view is the "ecological" way Darwin thought about his plants. He did not deny that much of Europe had been submerged in the more recent geological periods, but saw that there was no need for water to lap around the perimeters of depressed mountain ranges, so long as all the plants under discussion moved and acted as a single group. His explanation worked without Forbes's mechanical devices simply because he thought of arctic and temperate plants as coherent units that could, if required, form barriers as effectively as could geological phenomena. Darwin's arctic plants were trapped on the tops of mountains by a necklace of temperate species, and one must suppose that if the latter were removed arctic species would press outwards as far as they could, depending on the surrounding conditions. Forbes's arctic plants, by contrast, had been lifted to the heights at which they are now found and were in that respect somewhat passive entities. Forbes envisaged boreal species as organisms at the mercy of external mechanical agents—elevation and depression of land being the major factor here. Yet Darwin's thesis worked without these extra assumptions; he evidently had a gift for paring explanations down to the simplest possible hypothesis.

On the other hand, the similarities between Forbes's and Darwin's theories are more striking than the differences. Both men supposed a change in climate that allowed species of other latitudes to overrun what was left of Europe, and Darwin depended on snow and ice as a limiting factor where Forbes preferred a sea. With a subsequent change in temperature, both men suggested that arctic species would follow their native climates and return to northern latitudes or to the elevations which suited them best. The most striking similarity of all was that they both attempted to answer exactly the same question: why there were identical species in two dislocated spots. Darwin and Forbes saw this question in an almost identical light, even to its deeper meaning for the origin of species. Their explanations clearly refuted the doctrine of multiple creation. With so much in common it is plain that any discrepancies between the two theories are indeed minor.

Together, these explanations for the origin of disjunct species mark the establishment of dynamic distribution studies. An account of sporadic distribution patterns that also satisfied Forbes's and Darwin's requirements for single creation was to be found only by supposing that plants and animals have the gift of movement. Species did not simply remain in the general area of their origin; they could move, far and wide, and did. Species, suggested Forbes and Darwin, could disperse over the whole world if given enough time and appropriate mechanical means. Floras and faunas ebbed and flowed over continuous country, leaving small "pools" of species behind them or adding one further "layer" of organisms to an already multilayered population. Everything was flux and movement. The intellectual scope of such ideas was manifest.

DYNAMIC DISTRIBUTION

Darwin went on to expand his ideas about the movements of floras in more ways than one. He saw no reason to doubt that mountain species of North America and Asia were remnants of the same arctic vegetation that had populated Europe. Such a view explained similarities between the montane floras of Labrador, the Altai, the Himalayas, and the mountains of Turkey and Afghanistan. The theory was taken further in 1844, when he proposed that Antarctic plants, and those of upland New Zealand and South America, were also of northern origin.[30] Assuming that the cold period had affected the whole earth (a point over which Darwin laboured, conceding that it did not have to become cool all at once and to the same degree), he suggested that arctic plants had moved south through the tropics along suitably cold mountain ridges. The backbone of the Americas formed one such "highroad," enabling forty or fifty plants from Alaska or thereabouts to migrate along the Rockies and the Cordillera de los Andes to a final resting place in Tierra del Fuego. When warmer temperatures returned, these species would be cut off from their fellows and appear, as in the case of Sweden and Switzerland, to be disjunct.

Darwin was convinced that this dramatic event had taken place. He noted that plants such as the Antarctic beech were identical to organisms found in the arctic circle. The two populations must consist of one and the same species, the only alternative explanation being one of multiple creation. Moreover, there were other species not precisely identical but very closely allied among the floras of north and south. Darwin thought these were derived from a common ancestor or, more probably, that the southern form was derived from the northern through the action of natural selection on species far removed from their natural home. In Darwin's eyes the existence of such "representative" species provided strong supporting evidence for an earlier extension of a northern flora.

Darwin's work on arctic and alpine vegetation illustrated the immense distances plants could travel. Expanding on this, he went on in 1856 to devise a scheme of distribution that took the entire world as its chessboard.

Concentrating again on the effects of cooling temperatures, Darwin proposed that arctic species could cross the equator only where there was high land, such as the Andes or Himalayas. But he thought it entirely possible that temperate species could have penetrated and held their own in the lower lands between the tropics. Citing all the evidence he could muster to prove that the tropics must once have been less extensive and a shade cooler than the present day, he suggested that the tropical plants of the Pleistocene were under some strain and less likely to ward off invading temperate plants. "Tropical productions," he asserted, "would be in a somewhat distressed condition, and therefore would not oppose so bold a front, as before or subsequently, to the intrusion of strangers."[31] This view would explain why European temperate plants existed in Australia or the Cape and why species belonging to genera typical of the arctic circle appeared in mountainous New Zealand: they had migrated through the tropics to take up residence in southern countries. Acting as a sieve, the equatorial flora allowed only a few temperate species through into the south, and these began a "new" existence as members of other distinctly austral floras, many of them subsequently losing their northern identity through the influence of natural selection.

Darwin's theory was one of great explanatory power and flexibility. By these means, he said, "we can understand the presence of so many European forms in Tierra del Fuego, as is so forcibly shown to be the case by Dr. Hooker in the *Flora Antarctica*, some absolutely identical, some presenting strongly marked varieties and some quite distinct, but still plainly related to their northern congeners."[32] Here there was no room for multiple creation. Everything could be explained as the result of the movements of floras in concert with changes in climate and topography. It was an argument that required time, of course, but time was readily available in the Lyellian universe. If the conditions were right, species could move all over the earth in great waves, ebbing and flowing as the climate shifted from warm to cool to warm again.

The imagery of a "sea" of species evidently appealed to Darwin, who used an extended metaphor of the "living waters" of plants to close his account of global distribution in the *Origin of Species*. It reinforced the general impression that montane floras and faunas were insular in character, in itself a powerful metaphor for many distributionists of the period. Latreille, for instance, remarked on the striking resemblance between the insect life of the Alps and that of oceanic islands: in both locations beetles were adapted to their surroundings by being wingless and predaceous. It was easy to think of montane species as the inhabitants of an island and so, presumably, to think of organisms of the lowlands as "living waters." According to Darwin, these waters rose in a great tide of organisms, depositing species on the "land" before retiring again. Oddly enough, his words echoed Forbes's emphasis on water and may indeed have been derived from Forbes's views on the distribution of British plants and animals (Darwin is known to have reread Forbes's 1846 paper before composing his extended argument of 1856).[33]

Just as interesting as this watery metaphor is Darwin's stress on the "power" and "dominance" of northern forms of life over those of the tropics and the southern hemisphere. The idea seemed to be vindicated by several notable facts. Species of European origins—the point of origin being deduced from the large number of species in relation to genera—were more often found in southern regions than plants and animals of the south were found in the north. On Darwin's argument the earth had cooled sufficiently to allow arctic plants to migrate across the tropics; should not the Antarctic flora have done the same thing in reverse? It seemed as if the southern species were, in some way, weaker than those of the north. The same logic applied to the temperate floras. There was no evidence of a south-to-north movement beyond the existence of a few austral plants on the mountains of Borneo.

Darwin explained the imbalance between north and south by supposing there were more species and more individuals in the north. Statistically, these had a greater chance of penetrating the tropics, especially if the cold came on from the north. Greater numbers, he thought, were a direct result of larger expanses of land in the northern hemisphere. Moreover, under Darwin's mature theory of evolution by natural selection, these northern forms would be subject to more competition, resulting in "better" species that were genuinely "stronger" and more "dominant" than the "less powerful" southern forms. Telling evidence for his supposition came from a general consideration of introduced and naturalised species: although many European forms were readily naturalised in countries such as Australia, Darwin could not think of any southern plant or animal living in Europe except those in botanical or zoological gardens. Despite extensive commercial interaction between these countries, including the deliberate or inadvertent transport of species from one land to another, the flow of organisms seemed to be one-way only.[34]

The idea of a "dominant" northern flora formed an important part of post-Darwinian theories of distribution, reaching its apogee in the later works of Joseph Hooker, A. R. Wallace, and, to some extent, Louis Agassiz. These naturalists were not afraid to draw a parallel between organisms and men; they saw the predominance of northern plants and animals as a reflection of the general excellence of northern species of humankind, of the sophistication of northern societies, of the efficiency of northern economic policies. For them, the dominance of Europeans over men of other countries was an accepted fact. The annexation of colonies, the apparently successful administration of those lands, and the development of their natural resources for the benefit of the mother country all betokened the superiority of northern races. In many ways this opinion antedated evolutionary theory, but it received its greatest publicity when backed by the "laws" of natural selection. Hooker, for example, confidently assumed that it was not the meek but the Europeans who would inherit the earth. During the last third of the nineteenth century he tacitly endorsed the British settlers' barbaric behaviour in New Zealand, by suggesting that the aboriginal Maori inhabitants were destined to be exterminated by superior races. Antipodean men, plants, and animals would eventually give way, under the rules of selection, to "better" organisms originating in the north.[35]

There is little in Darwin's printed or private papers to indicate what, if anything, he thought about the parallel between northern man and northern animals and plants. In the *Origin* and elsewhere he firmly avoided discussing anything other than animals and plants. Nonetheless, there was a definite feeling in early Victorian Britain that the Europeans' innate skills made them naturally preeminent in world affairs. However indirectly, Darwin's insistence on the hereditary vigour and power of northern species of plants reflects this feeling.

HOOKER'S COLLABORATION

Although Darwin's friends were rarely asked to read his preliminary sketches of natural selection theory, Joseph Hooker (1817–1911) was one of the chosen few and the naturalist to whom Darwin most frequently turned for advice and criticism. In 1856 he was invited to comment on Darwin's case for arctic-alpine floras in a piece being prepared for his "big species book." It was the first sample of Darwin's work that Hooker had read since he had gone over the short *Essay* of 1844 at Darwin's own request. No doubt he was flattered to be singled out in this way. Darwin knew Hooker to be the most suitable "sounding-board" for such a theory and lost little time in asking him to read the paper as soon as it was finished. By 30 July 1856 however, the manuscript was still not copied out and Darwin had to wait impatiently for Hooker's verdict until the latter returned from a summer walking tour in late September. When Darwin's forty or so pages eventually reached their destination, Hooker found that the author had added a note exhorting him not to be "too severe," nor yet "too indulgent." Reading between the lines it was evident that Darwin was anxious for Hooker's approval.[36] The piece was primary material for showing that multiple creation was a fallacy.

Darwin's thesis of dynamic floras and faunas greatly interested Hooker. He too was a disciple of Lyell and was ready to welcome a combined geologicobiological scheme, especially one based expressly on the notion of migration after a single creation. The arctic-alpine part of Darwin's theory clearly refuted multiple creation and provided an admirable model for the history of distribution as waves of floras or faunas moving over the earth while climates changed and surface configurations altered. Yet Hooker expressed concern at what appeared to be an important omission in Darwin's argument: he could not see how Darwin (or, for that matter, Edward Forbes) proposed to transfer species from one region to another, from point A to point B. Representing the views of more conservative and traditional naturalists, and reflecting his own exacting standards of botanical explanation, Hooker was not satisfied with the mechanics of dynamic distribution.

The suggestions both Darwin and Forbes put forward for the dispersal of plants seemed patently inadequate to Hooker and in some cases entirely against the facts of present-day distribution. To characterise each position briefly, Forbes had presented a scheme in which the "island in a glacial sea" constituted the major theme, with arctic plants being transported from polar regions by means of icebergs; Darwin, who had not written of seas or islands, was also forced to rely on icebergs to breach the oceanic portions of the globe that had undoubtedly existed during the Pleistocene; Hooker, on the other hand, preferred a land bridge over which arctic plants could have migrated with no further help.

Somewhat unjustly under the circumstances, Hooker accused Darwin of ignoring the question of dispersal. By 1856 Darwin had conducted many experiments on possible modes of dispersal, details of which were to appear in the *Origin* to substantiate his view that organisms can migrate over vast distances and against apparently insuperable odds; these points were yet another aspect of his argument against multiple creation. But Darwin did not include this section of his work in the manuscript sent to Hooker. It seemed to Hooker, therefore, that Darwin had treated the whole question in extremely cavalier fashion.

In fact there were many practical reasons for Hooker to reject the idea of iceberg transportation. He had spent years circumnavigating the southern ice pack with Captain Ross on the *Erebus* and *Terror* and had rarely seen ice floes carrying sticks, soils, or plants, even though he had been looking for exactly that phenomenon (as he told his family back home and his father's friend, the elder Charles Lyell).[37] If icebergs were to be introduced despite this objection then at least they should be introduced properly: neither Darwin nor Forbes seemed to realise that icebergs travel in more directions than simply north-to-south. Currents would sweep many icebergs in an east-west direction, and, of course, in the Antarctic seas ice would travel most frequently toward the north. If Forbes and Darwin were to propose floating ice as the prime dispersal mechanism then they must be prepared to explain why most species seemed to have diffused from the north pole, and along lines of longitude.

In addition Hooker noted that such irregular, accidental methods of dispersal would give rise to a typically "insular" set of floral relationships, which the arctic elements of the Alps, Australasia, and South America did not exhibit. Because islands were usually colonised by an odd assortment of species, one from one family, one from another, the proportion of species to genus was notably extreme. Tenerife, for example, possessed roughly three species for every two genera on the island. Coral islets were particularly remarkable in this respect; although poor in plants they usually exhibited high diversity in the families making up the flora. To Hooker, iceberg transport seemed just as random and accidental as oceanic island colonisation. But existing arctic and arcticlike floras did not exhibit these characteristic proportions: there were many species in each genus of *Silene* or *Saxifrage*, for instance. Hooker therefore reasoned they must have diffused over dry land, since the flora could not have retained the original associations of plants in any other way.

In his letters to Darwin and in some expansive footnotes to his 1855 essay introducing the New Zealand flora, Hooker sketched out his own preferred scheme, which he later published in an important paper, "Outlines of the distribution of Arctic plants," read to the Linnean Society in June 1860.[38] A simple change in isotherms, he suggested, would alter the European climate to something like that of Labrador. According to Lyellian theories, this could be effected by the elevation of a great continent between eastern America and northern Europe, which would lower the average winter temperature while making the climate more extreme. This idea dispensed with iceberg transportation and gave arctic plants dry land for the migration from Sweden to Scotland, or from Lapland to Labrador, and thence to the south along Darwin's "highroads." The same general explanation accounted for the distribution of Antarctic plants, where Hooker supposed a great southern landmass connecting the south pole with Australasia and, at some other time, the pole with Chile and Tierra del Fuego.[39] The elevation of such a continent would, under Lyell's theory, provide a change in climate allowing Antarctic plants to spread outwards. A submersion would provide a warming-up period as temperatures returned to their former levels. The flora left behind in Australia, New Zealand, and southern America would therefore be rendered disjunct.

Hooker was evidently so taken with the possibilities of a dynamic interpretation of earth and life, and with the correlation of geological and organic movement promulgated in Lyell's, Darwin's, and Forbes's work, that he tried to refine his colleagues' arguments with a dynamic version of his own. If nothing else, these variations on a major theme illustrate the potency of Lyell's ideas and the rich theories that could be derived from them. Hooker believed these views could explain the distribution of species—even the difficult problem of disjunct species—on the simple assumption of movement from an earlier location. The work of Forbes and Darwin therefore stood as powerful testimony to the origination of species on one spot, from which they dispersed according to the vagaries of the earth's history. The idea was so appealing that when faced with practical objections, such as the inadequacy of Darwin's mode of transport for plants, Hooker preferred to add a few land bridges rather than relinquish the general argument. He saw Darwin's writings as the vindication of the theory of single creation.

THE PERMANENCE OF TYPE

But Hooker was puzzled about Darwin's references to "representative" species. If species A travelled all the way from England to Tasmania and then turned into species B, how could one ever verify the theory that it arose in one spot? There was no specific identity between the two, so the explanation used for disjunct species did not apply. Yet Darwin seemed to be arguing to the contrary.

Hooker asked Darwin to explain his transtropical migration in more detail. Since the temperate regions were closer to the equator, surely temperate plants would have begun to migrate before those of the arctic? And if they had had longer to colonise the southern hemisphere, why were so few of them to be found there?[40]

The answer to Hooker's query lay in a complex piece of Darwinian reasoning. Certainly the temperate plants would have moved as soon as the cold came on. Certainly they penetrated the tropics long before the arctic plants. But, on the theory of natural selection, these first movers were inevitably exposed to "new associates and new conditions" and hence subject to change; most of Darwin's temperate migrants would develop a different form from their northern originals and could more properly be called "representative" species. There were very few absolutely identical temperate species on both sides of the equator because the southern ones had all changed. By implication, since the migration was a continuing movement, the species Darwin called the "last" to cross the tropics were exposed to new conditions for the shortest period of time. Consequently, these were less likely to change. Darwin's argument thus accounted for the fact that most arctic plants had remained the same throughout their global travels whereas temperate species tended to change their form when relocated in the southern hemisphere. "Time," he told Hooker, "is a most important element."[41]

But Hooker spotted a flaw in Darwin's reasoning and riposted in no uncertain terms. Since time was the crucial element, why had not the arctic forms changed more than the temperate ones? They had twice as far to go and took twice as long to do so; surely they ought to have become modified? Darwin's hypothesis failed to account for the fact that arctic plants were exposed to more fluctuations than most in their global travels. "You expose them to more extreme conditions than any other plants can ever be subjected to," he complained.[42] Should they not change too? As Hooker saw it, Darwin was arguing, on the one hand, that time gave altered conditions and new associations leading to changes in temperate species and, on the other, that the same elements led to *no* modification in the case of arctic plants. "I think this will hardly hold water," he told Darwin.

Yet Hooker had failed to grasp Darwin's ideas of time and change in species. For Hooker, the doctrine of the origination of species in a single location required a certain permanence of type. If species began with only two members they must have retained their specific characters over a long period of time and through a logarithmic increase in numbers, otherwise the same species would not be in existence today. This was the basis for the Darwin-Forbes explanation of the origins of disjunct species. Yet Darwin was also mixing in a certain amount of transmutation and saying that over long periods some species were liable to change. Hooker, though "never very stubborn about the unalterability of specific type,"[43] disputed the logic of Darwin's double-sided view of time: Darwin really ought to decide if the passage of time was, or was not, marked by organic change. He should not conflate the two in the same theory.

But Darwin saw time's flow as being in some way independent of the organic and inorganic changes by which it is usually measured. Neither the physical constitution of species nor the physiographical construction of the earth were, for him, necessary indicators of the passage of time. He seemed to believe that the hands of a clock, for instance, were a mere convenience to give men some idea of the passing of another absolute time, an idea which Newton had also pointed out nearly two hundred years before. "Time," Darwin wrote, "is important only so far as giving scope to selection."[44] In this way he could claim that some species remained the same over vast periods whereas others changed. The lapse of more and more "parcels" of geological time simply made it increasingly likely that there would be alterations in the earth's crust and a rearrangement of species; *these* were the operative factors in stimulating organic change.

Hooker was thoroughly perplexed. This was not the sort of time he understood, where the passage of years could be gauged only by the change from one state of affairs to the next.

But he evidently admitted the importance of Darwin's view. Hooker confessed to having "never felt so shaky about species before."[45] The theories drawn out by Darwin did more, in his opinion, than account for disjunct species in terms of single creation; they hit right at the heart of the problem of creation itself. Had Darwin found the answer to the species question?

SIX

A Keystone to Creation

Although Edward Forbes had proposed a perfectly acceptable hypothesis for the origin of disjunct species, together with strong arguments for a theory of single creation, these brought him no closer to revealing how species first appeared in nature. He could explain peculiar distribution patterns like those of the strawberry tree (*Arbutus unedo*) by supposing the Asturian flora to have extended its boundaries from northern Spain to southwest Ireland, but the question of the first appearance of Asturian plants remained unsolved. How had these and other species been called into existence? As Sir John Herschel pointed out in 1830, this was the greatest mystery of natural philosophy.

Much of early-nineteenth-century research in the earth and life sciences was directed toward the solution of the so-called species problem and the actual mode of organic creation. Information from physiology, embryology, comparative anatomy, systematics, palaeontology, and the experimental study of hybrid organisms was widely used in a search for rules and regularities behind the phenomena of nature. It was expected that such rules, once discovered, would shed light on the origin of species.

In this *mêlée* of intellectual activity geographical distribution did not pass unnoticed. On the one hand, it was a subject dedicated to patterns and processes in the organic world, to the charting of regularities in the topographical and chronological arrangement of species. On the other hand, it supplied data which, to be understood, required the formulation of hypotheses about the first appearance of living beings. Many naturalists were convinced that qualities such as these should not be ignored and enthusiastically made the study of distribution an important feature of their search for creation's laws. "The subject," declared Darwin in 1845, was a "key-stone to the laws of Creation."[1] His belief in the significance of geographical distribution was echoed by other philosophical naturalists, among them Forbes and Louis Agassiz. "One great thought prevails in natural historical studies," wrote Agassiz in 1842, "the study of the laws regulating the geographical distribution of natural families of animals and plants upon the whole surface of the globe."[2] Generalisations from botanical and zoological geography were of the highest interest to naturalists searching for a deeper understanding of nature.

One phenomenon in particular inspired a wealth of hypotheses pertaining to the origin of species. This was the existence of what were then called "representative" organisms: pairs or triplets of closely related species which individually occupied different geographical areas. They were similar in structure but mutually exclusive in topographical location. Superficially, the differences in specific form seemed attributable to each living in a separate area. But how had such differences been acquired? Through original creation, transmutation, degeneration, variation, or isolation? For many naturalists representative species seemed to hold the clue to the story of creation itself.

Interest in the problem ran high. Representative species took the centre of the stage, as it were, from the early 1830s through to the end of the half-century, and the focus of geographical research shifted toward a closer examination of their origins and ultimate significance for science. In the first instance, explanations were dominated by the wish to understand the obvious phenomenon of topographic replacement. But the issue was complicated by abstract speculations about the meaning of analogy in nature, coupled with anxiety over the classificatory status of such organisms. Overall, an unwieldy collection of ideas, facts, and assumptions about representative species had come into being. The question was, how were these issues to be resolved? While the botanist Joseph Hooker had serious doubts about the possibility of any satisfactory solution, Agassiz and Forbes thought they knew the answer. Together the three of them presented a rich diversity of theories relating to the creative processes of nature.

AGASSIZ AND MULTIPLE CREATION

Louis Agassiz's (1807–1873) ideas are of enduring interest to historians, if only for the fact that they always seem to be the opposite of whatever theory was currently in favour. His works carry the attractive aura of being written by a troublemaker, a thorn in the side, acting as a goad to less thoughtful and knowledgeable naturalists. Sometimes this eccentric position worked to his advantage. He was sufficiently divorced from mainstream geological practice, for instance, to see in the structure of the Alps, and then in the Grampians of Scotland, evidence for extensive glaciation, which led to the proposal in 1839 of a northern ice-sheet during the Pleistocene period. British geologists, favouring a familiar and flexible theory of extensive oceans bearing floating ice, rejected Agassiz's *Eiszeit*, only to accept it with some adjustments during the 1860s.[3] In this case he was proved right. But his maverick views on the multiple creation of species, and the "autochthonal" origination of all individuals, all over their range, all at once, were never to be accepted, either then or now. Agassiz was so sure of his own opinion here that he argued himself into an ideological corner impossible to defend. Although there was considerable anti-Darwinian feeling in American society during the 1860s, and although Agassiz issued many polemics criticising evolution (in general) and descent from an ancestral pair (in particular), neither his fame nor his creationist stance were sufficient to inspire any followers: his theories were too extreme, even for those who rejected Darwin. Only Asa Gray, his colleague at Harvard, seems to have taken these suggestions at all seriously, discussing them at some length in his *Darwiniana* and in occasional reviews.

Nonetheless, during Agassiz's years in Europe and his first decade as professor of comparative anatomy at Harvard, his views contributed in no small way to the development of palaeontology and animal distribution studies. From his arrival in America in 1846 to the publication of his *Essay on Classification* in 1857 and beyond, Agassiz was widely respected. In the uncertain biological world of the second quarter of the century, no one could say that his suggestions were right or wrong, sensible or foolish. It was evident that he thought deeply about the philosophical assumptions behind contemporary biology, and clear that he brought an excellent mind to bear on the issues of the day. So Agassiz's ideas about species were certain to be read and carefully considered; his contributions, though unusual, were part of the same biological debate as those of more conventional naturalists.

The most trenchant statement of Agassiz's theories appears in his *Essay on Classification*, preface to an unfinished *Contributions to the Natural History of the United States* (1857).[4] He stoutly defended the idea that living beings were independent of physical influences. To him, life was the expression of a higher intelligence, where the divine plan was revealed by the structure and interrelations of His species. Agassiz based his defence on the belief that physical forces were inadequate to explain the complexity and diversity of the organic world: the uniform action of, say, temperature could not produce the multifarious properties of species, let alone the more abstract qualities of the higher taxa. Nor could soil or atmosphere influence the intricate relations between organisms—relations of behaviour, frequency, gregariousness, and so on. Important supporting evidence came from his own palaeontological and anatomical work, which established beyond doubt that recently discovered forms of life invariably possessed characters relating them to species already known to science, so demonstrating, first, a plan for nature and, second, the existence of many hidden and unexpected links between living beings.[5] This evidence convinced Agassiz that life consisted of diverse, principally unknown, relations between its units. It followed that such a complex organic web could only have been created all at once, as it now stood, every species in its place with its various connections already in working order. The network must have been preordained, he insisted, and the entire fauna created simultaneously.

These assumptions were essential to Agassiz's larger view of things. He believed the earth had undergone successive "revolutions," during which its animal life had been periodically wiped out and renewed in accordance with major changes of climate and topography. No organisms escaped extinction at the end of an epoch, because the delicate interrelationships between them and their physical environment were utterly destroyed. Accordingly, all nature had to be re-created at the beginning of the next geological epoch. And each successive re-creation, argued Agassiz, introduced new, more complex animals and plants: living beings had increased in "perfection" through the ages, vertebrates in particular having advanced in a steplike sequence toward a pinnacle of development culminating in mankind. In this way, the organisms of every major era were linked by comparative anatomy but separated by the unmistakable barrier of geological revolution. More significantly for Agassiz, the reigns of beast and man were completely sundered by the last and most sensational revolution of all—his dramatic postulate of an Ice Age.

The question of the repopulation of the earth after each major extinction led Agassiz to suppose a multiple origin for species. If animals were introduced through piecemeal single creation, part of the earth, he thought, would be temporarily understocked with living beings: it would take a long time for the new breeding couples to multiply and disperse over the unpopulated globe. The fossil record revealed no such gap to Agassiz. Even allowing time for species to increase and spread, what then of Agassiz's concept of an integrated and intradependent ecosystem? Palaeontological researches told him that the arrangement of living beings in each geological period was every bit as complex as it is now; surely these fossil faunas were too complicated to be the result of piecemeal creation followed by accidental dispersal? "The details of the geographical distribution of animals," he wrote, "exhibit, indeed, too much discrimination to admit for a moment that it could be the result of accident, that is, the result of the accidental migration of the animals or of the accidental dispersal of the seeds of plants."[6]

It was much more sensible, he claimed in 1857, to suppose that species originated "within the natural boundaries they occupy [today]." The assumption of an aboriginal pair of individuals for each species could be abandoned in favour of an "autochthonal" theory where everything was created together. "Pines have originated in forests," he stated; "heaths in heathers, grasses in prairies, bees in hives, herrings in schools, buffaloes in herds, men in nations!"[7]

This doctrine of multiple creation absolutely precluded any modification of species. The slightest alteration to the finely tuned network of interrelations would throw the whole system into disrepair. For Agassiz, there could be no "progression" or advancement of the organisation of species inside each geological time span. Nor could there be any additional creative acts during the course of one period, since this too would alter the original balance of species. His appreciation of the complex nature of the ecosystem dictated that it must always be stable. Change in any of the components would only lead to its collapse. From this idea it followed that all species found in the formations and strata of each major geological system had been created in one giant batch, an unusual view that seemed to ignore the great lengths of time encompassed in a single geological system, the diversity of conditions indicated by changing strata, and the variety of animals and plants fossilised therein.

It is evident, then, that Agassiz's philosophy of nature relied heavily on the notion of hidden complexities and inbuilt correlations between organisms and on the idea of an internal integration for faunas. These he derived from a study of animal distribution and, in particular, from examining the phenomenon of representative species.

Who, asked Agassiz in his *Essay on Classification*, could look at the faunas of North America and western Europe and fail to note that some species were the same and others different? If the two populations had been exactly the same or completely different, he conceded, there would be little difficulty in accounting for them under current hypotheses of specific origin. But, like Buffon, he was fascinated by the paradoxical interplay between the two: the transatlantic faunas formed neither a single biological province (since each continent possessed indigenous animals of its own), nor were they two separate provinces (since they shared a large number of species). Could single creation account for such facts? Agassiz thought not and drew upon representative species of ducks to expound his case.

The widgeon and the redhead duck each have two distinct forms, confined to either side of the Atlantic, whereas the mallard and the scaup are the same throughout North America and Europe. There was no question in Agassiz's mind that the two widgeons, for example, were separate species. Taxonomic differences were strongly marked and held true through many generations. He could not allow that environmental factors had caused these representative forms to diverge in character from some common ancestor, because the same influence produced no comparable effect on the mallard or on the scaup. Two kinds of widgeon, as both single and multiple creationists agreed, must have come into existence, each in its own geographical area. The problem, as Agassiz explained, was to square this idea with the appearance of a single mallard in *two* countries. Had the mallard originated in one or the other region and then spread across the Atlantic? Since it lived happily in both countries, which was its original home? Moreover, as Agassiz pointed out, ducks tend to range from north to south, not from east to west. He could not believe that a species created as a member of one fauna could successfully infiltrate and maintain itself in another. Perhaps, then, the whole transatlantic fauna had been created in one continent, only to rearrange itself in this peculiarly asymmetrical fashion? But why, under this scheme, would there be a need for two species of widgeon? And where were the relics of American species in Europe and of European species in America that this theory would suggest? Surely, asked Agassiz ironically, there could not have been a mid-Atlantic landmass on which the fauna appeared and from which it subsequently disbanded, some species to the east and others to the west, all during the recent period?

It was obvious to him that representative species held the answer. Here were two closely allied species of widgeon with separate places of origin. If one supposed a dual creation for widgeons, why not for mallards too? Under Agassiz's interpretation of nature there was no other feasible suggestion. And if the Creator made four original mallards, two for each country, why not many hundreds of mallards, the entire roll call of ducks, all in one giant flock?

> By way of an example, I would mention the European and the American Widgeon . . . or the American and the European Red-headed Ducks . . . which inhabit respectively the northern parts of the Old and New World in summer and migrate further south in these same continents during winter, while the Mallard and the Scaup Duck are as common in North America as in Europe. What do these facts tell: that all these birds originated together somewhere, where they no longer occur, to establish themselves in the end within the limits they now occupy?—or that they originated either in Europe or America, where, it is true, they do not live all together, but at least a part of them?—or that they really originated within the natural boundaries they occupy? I suppose with sensible readers I need only argue the conclusions flowing from the last supposition. If so, the American Widgeon and the American Red-headed Duck originated in America, and the European Widgeon and the European Red-headed Duck in Europe. But what of the Mallard and Scaup, which are equally common upon the two continents; did they first appear in Europe, or in America, or simultaneously upon the two continents? Without entering into further details, as I have only desired to lay clearly a distinct case before my readers, from which the character of the argument, which applies to the whole animal kingdom, may be fully understood, I say that the facts lead, step by step, to the inference that such birds as the Mallard and the Scaup originated simultaneously and separately in Europe and America, and that all animals originated in vast numbers, indeed, in the average number characteristic of their species, over the whole of the geographical area.[8]

Here, Agassiz used representative species to promote concepts entirely at odds with the ideas of Lyell, Forbes, and Hooker. He saw these species as organisms providing each flora and fauna with a unique and unchanging character. Representative species were a "striking proof" that "all animals and plants have occupied from the beginning, those natural boundaries within which they stand to one another in such harmonious relations."[9] Between Agassiz's groups of species there was no dynamic interaction of the kind prescribed by Darwin and Forbes for alpine and arctic floras. They were instead static units, unable to change the internal proportions of this family or that, incapable of accepting or donating member species to other associations, and powerless to change their boundaries over time. There had been no past existence for Agassiz's faunas, no history beyond the fact of their creation. The time span between that moment of creation and the present day had not been filled with interesting changes and fluctuations, nor had it documented the multiplication and dispersal of species from their original parents. Agassiz, despite his keen knowledge of palaeontology and an abiding interest in determining the sequence of life on earth, was in essence a "pattern" man who could not discern the dynamics of populations. Representative species served to illustrate a theory of nature that was, at root, fixed and predetermined by some higher being.

REPRESENTATION IN SPACE

Edward Forbes's views on representative species, by contrast, were almost totally concerned with sequence. The appearance of geographically exclusive forms was interpreted as part of a larger theory of multiple analogies or "representations" in the natural world. Various elements of this underlying theory of nature were described in Forbes's successive papers on distribution, one significant aspect being labelled the "principle of polarity" and presented to a largely uncomprehending audience at the Geological Society in 1854. The entire doctrine was never given to the public as a coherent whole and it has to be pieced together from references in his works and correspondence to a "system" which had dominated his thoughts from his student days at Edinburgh and which formed the motivation behind his greatest conceptual achievements. Forbes hoped to discover the phenomenon of representation in the different dimensions of space (area), time, height, and depth. "Respecting the doctrine of representative species," he confided to Leonard Horner in 1845, "I am very anxious you should understand it, as it is one I lay great stress upon and believe that on its development the fixing of Geology as a science will eventually in part depend."[10]

The first facet of Forbes's theory was put forward in early papers on marine biology and later expanded in his work on the mollusca and radiata of the Aegean seas.[11] He took pains to demonstrate that a "replacement" of one set of species by others took place according to the depth of the ocean floor. Dividing the seabed into zones or belts, just as Humboldt had done for plants at differing elevations, he observed that each belt carried a distinctive fauna and flora. The overall character of these associations depended on the depth at which they lived, becoming more arctic or boreal in nature as the depth increased. Forbes had been struck by the discovery in 1841 of molluscs in the deeper reaches of the Aegean that belonged more properly to the Celtic marine fauna. To pull up Irish or Scottish species while working under a Mediterranean sun must indeed have been perplexing. These strangers, he later wrote, were like "the Basques among the Spaniards, or the Cornish among Englishmen, relics of ancient possessors of the country whose epoch of dominance has ceased to be, but [which] still remain in fragmentary masses, as if to show what and where they once were."[12]

But, as Forbes was well aware, this was a relationship of biological similarity rather than identity. "In each case the analogy is maintained, not by identical species only, but mainly by representative forms." The Mediterranean species dredged from below 105 fathoms (1,380 feet), the upper limit of Forbes's eighth and final zone, were not necessarily identical to those he knew so well from Scottish waters, but were closely allied species, often members of the same genus or more exceptionally belonging to allied genera. "The forms of representative species," he insisted, "are similar, often only to be distinguished by critical examination."[13]

Like Agassiz, Forbes believed implicitly in a divine origin for all living beings and that the variations between such representative forms were consequently part of the special identity given to each at its creation. But, unlike Agassiz, he interpreted past and present distribution patterns as the result of single creation. That act was necessarily followed by a gradual increase in the numbers and geographical range of each species, probably localised in time and space, so that the first members of a single flora or fauna came into existence at roughly the same time and in approximately the same area, thereby giving the impression that each had a definite "point of origin." Every flora or fauna, Forbes supposed, had been created in a form ideally suited to the existing conditions of the earth. As he went on to show in 1846, he thought these groups of species could expand in range just so far as that particular combination of conditions prevailed. Upon such an assumption he based his explanation of the reappearance of disjunct species in two or more widely separated regions.[14]

Yet this view was hardly enough to explain the existence of representative species. By definition these were not the same but rather analogous or closely similar species. A purely mechanical device such as that used by Forbes to account for arctic-alpine organisms would not be valid, precisely because the species were not identical. Some creative process must have intervened. This much was clear to anyone who believed in a divine origin for each and every species. But why should they have been created so similar in form?

It appears that Forbes believed each of his representative species originated in a different flora or fauna. He postulated that the Pleistocene, for example, was populated by at least two plant communities, one of which originated in Scandinavia, the other in the greater Mediterranean area, and which both contained species adapted to the "glacial" conditions of the age. Because in his view the same species could not be created twice—even if the conditions were strikingly similar—Forbes suggested that the Creator had given the members of this pair of floras some minor differentiating characters. The same argument applied to Pleistocene animals: shells dredged from the depths of the Aegean were indeed analogous to or representative of those found off the shores of Scotland; they were alike because they had been created for the same kind of environment, at a time when the whole of Europe was subject to approximately the same external conditions; they were different because the particular combination of characters that make up one specific form were, by definition, irrepeatable.

This was the line Forbes took in his 1846 paper "On the connexion, etc." To the example of marine molluscs he added the case of Swedish and Alpine floras. Although these two plant communities possessed many species in common, each also contained other species special to their own region, such as the gentians of Switzerland. To explain this Forbes proposed two separate creative acts. If some of these endemic organisms happened to be very similar in form to members of the other flora, it was because they had been created for the same conditions. Through the agency of the Creator, the species mimicked each other.

In a few terse sentences Forbes emphasised that Humboldt's law of the parallel between belts of latitude and elevation did not depend exclusively on the reappearance of precisely the same species. More significant was the maintenance of the parallel even in the absence of disjunct species. It depended on the existence of species that were merely similar to each other: it depended on representative forms.

> Tournefort first remarked, and Humboldt, the great organiser of the science of natural-history geography, demonstrated that zones of elevation on mountains correspond to parallels of latitude, the higher with the more northern or southern, as the case might be. It is well known that this correspondence is recognised in the general *facies* of the flora and fauna, dependent on generic correspondences, specific representations, and, in some cases, specific identities. But when announcing and illustrating the law that climatal zones of animal and vegetable life are mutually repeated or represented by elevation and latitude, naturalists have not hitherto sufficiently (if at all) distinguished between the evidence of that law, as exhibited by *representative* species and by *identical.* In reality, the former *essentially* depend on the law, the latter being an *accident* not necessarily dependent upon it, and which has hitherto not been accounted for.[15]

Forbes went on to outline his belief that similar analogies would be found between all floras and faunas created for the same conditions. Looking to the southern hemisphere, he claimed to see in Antarctic vegetation some species that mimicked those of the north. Here, though, the resemblance was not through representative *species* but through the appearance of analogous *genera,* as if increased topographical distance between floras was reflected in an increased taxonomic distinctness in their parallel organisms. To Forbes, this was an important and noteworthy phenomenon, indicating the underlying harmony of nature. If identical species and representative species existed, then why not further analogous forms, each presenting fewer and fewer features in common?

Here Forbes was picking up a generalisation expressed fairly often in the literature of the day, one which usually went much further than his own epigrammatic statements. It had long been a commonplace in distributional studies that species occupying adjacent areas were usually the same and those of more distant regions were likely to be different. Several naturalists, the eighteenth-century zoologist Zimmermann among them, had noted that this phenomenon was actually a gradient, where the identity between floras and faunas diminished according to distance, the classic example being Johann Forster's description of a diminution of Asiatic plants and an increase in American ones as he travelled eastward across the Pacific.[16] In the early years of the nineteenth century such gradients had been noticed more and more often and shown to appear in many different classes of living beings: H. C. Watson wrote of Scandinavian plants appearing in Scotland and then, in decreasing numbers, toward the southern parts of the Pennine Chain;[17] James Wilson did the same for insects, illustrating the differences between the entomology of the British Isles and continental Europe;[18] Theodore Cantor discussed the decrease of particular reptiles in Malaya according to degrees of latitude;[19] and William Thompson wrote on the absence in Ireland of certain birds prolific further to the east.[20] These facts were seen as a graduation of floras or faunas according to geographical proximity.

Such an interpretation encouraged the idea of a scale of resemblances between floras and faunas. Neighbouring regions ought to possess identical species, and distant ones different species. Intermediate associations ought to exhibit intermediate relationships. The entomologist Lacordaire, writing in 1842, put it this way:

> When two regions which resemble each other are at the same time contiguous, their species are almost wholly alike; this is witnessed, for example, in France and Germany. If these regions are separated by a considerable interval, their species are for the most part different, but the genera remain the same; it is thus that Spain and the coasts of Syria . . . possess a multitude of common genera with species almost precisely similar. Finally, if these regions are separated by an immense interval and by natural barriers, nothing continues common to them but the families. . . . It is this diversity produced by distance, which gives to each region its proper entomology.[21]

He connected an accumulation of organic diversity—or, more properly, morphological distinctness—with increasing topographical distance. The further apart two countries were, the more distinct the species.

For Forbes, and for another marine biologist, Henri Milne-Edwards, the important thing here was physical representation. Expanding the idea of scale above, they claimed that the further apart two countries were, the more distant must be the "representation" in taxonomic terms. Milne-Edwards drew an example by describing the distribution of freshwater crayfish in terms of a scale of decreasing morphological affinity:

> So far as the Crustaceae are concerned . . . many zoological regions exist, having each a peculiar population which is partly composed of organic types, whose analogues are not found elsewhere, and another part

> consists of the representatives of species which also exist in other regions. Thus our river crawfish does not exist in America, but is there replaced by a neighbouring species of the same genus; Southern Africa possesses a third species of the same animal and this organic type is also found in New Holland, but with distinct specific characters [that is, belonging to another genus].[22]

There was no question of a genetic or developmental connection between these differing species; they merely mirrored each other's structure to a greater or lesser degree. For pre-Darwinian naturalists, with no concept of evolutionary change, the relationship must have seemed part of the natural order of things. For Forbes, certainly, it indicated that the whole of nature could be understood by reference to the quality or extent of "representation" exhibited in living beings. It enabled the various relationships of plants and animals in topographical space to be ordered and set in sequence. Although the word as Forbes used it meant little more than, say, "analogy," he nevertheless elevated it to the status of a primal force in nature. For him, representation was the fundamental organising principle, the elementary, God-given cause for life's diversity.

THE "PRINCIPLE OF POLARITY"

By 1846 Forbes had declared his belief in a series of replacements or representations through topographical space. This spatial relationship of form with form, association with association, was important to him—not so much for the explanatory value it placed on his accounts of present-day distribution patterns, although this was no doubt useful, but for its value as a model of what went on in time. Just as replacement and representation could be found in space, so he believed there must be equivalent processes in the history of living beings. "As in space, so in time," said Lyell, a maxim to which Forbes fully subscribed. The relationships of species and genera over time could be explained by the relationships already discerned between floras and faunas of the present era. The rules of the game, as it were, ought to be the same for both.

Forbes went straight to the heart of geological theory. His views would not hold water unless he could prove that the history of life was made up from consecutive floras and faunas, just as the modern world presented consecutive communities in space. He attempted to demonstrate that the fossil record was indeed one of successive creations, that each batch of species had a distinct chronological boundary. Without such a boundary he could hardly claim that one set of species replaced an earlier set, an essential prerequisite for any analogy between them. So, in 1848, he put the geological question, "have genera, like species, centres of distribution?" Did genera occupy a "definite area in time?"[23] Forbes concluded that they did, citing an ad hominem argument that all known species appeared on the earth, continued for a certain length of time, were sometimes replaced or supplemented by others belonging to the same genus, and eventually became extinct. When the last species of a genus left the fossil record, the whole genus must be assumed to have been extinguished. If a species reappeared after a considerable lapse of time, the doctrine of single creation required geologists to interpret those organisms as survivors of an earlier genus, rather than a second creation of the same species. Relics in time could only be understood as chronological equivalents of disjunct species in space.

Some years later, when addressing the annual meeting of the Geological Society in London, Forbes still had not forgotten this crucial aspect of his system. He made great play with a monograph on the Brachiopoda (1851–1886), shallow-water marine bivalves, written by Thomas Davidson, which provided the detailed evidence his assertion needed. Davidson's "excellent work" demonstrated (to Forbes, at least) that each genus of brachiopods occupied a discrete "area" of time—a unique "generic time-area" as Forbes put it—analogous to the spatial limits of floras and faunas of the present day.[24]

Having made the point to his own satisfaction (although it stimulated the young Alfred Russel Wallace to write a strongly worded refutation from his temporary outpost in Sarawak), Forbes proceeded to unveil his theory of representation through time. This system, obscurely designated the "principle of polarity," was the *pièce de résistance* of Forbes's presidential address to the Geological Society of 1854, a philosophical gesture which caught his audience totally unawares. After its publication in the *Quarterly Journal* of the Society, Forbes went into further detail in a paper submitted to the Royal Institution. Since he died soon afterwards, both presentations were tactfully dismissed as the agitated ramblings of an unwell mind, and the principle of polarity was relegated to a no-man's-land of unacceptable geological theories.

The scheme was based on the Germanic, transcendental idea of the organic kingdom resembling the inorganic, with a "manifestation of

[a] force of development at opposite poles of an ideal sphere." Taking the entire history of living beings, Forbes proposed that the force of development could be seen at its most intense in the earliest and the most recent geological epochs, giving rise to a sort of north and south pole of the history of life.

Using his own idea that genera appear in the fossil record only for a limited span, Forbes suggested that these taxa came into existence in "batches." Each batch replaced the one preceding it, and through the structure of its species "represented" the outgoing set. To take an elementary example, he might have supposed that the Tertiary population of Great Britain possessed a species of turtle "representing" the Jurassic form of this reptile. But this did not necessarily mean a replacement of lower groups by something more advanced, for he was firmly against the doctrine of progression or development in which the sequence of living beings progressed toward a goal embodied in the structure of man. He claimed merely that one related being was replaced by another, as happened in the case of representative species in topographical space.[25]

But the important point for a doctrine of polarity was his assertion that the number of genera coming into existence at any one time varied from age to age. He believed he could find more new genera in the early formations of the Palaeozoic epoch than in the Mesozoic and, again, more new genera in the later formations of the Neozoic (Forbes's neologism for the most recent geological periods) than in the preceding age. Stressing that these statistics related only to innovation and to the introduction of *new* genera, Forbes here attempted to describe the birthrate of distinct sets of organisms. The rate was thought to be significantly higher in certain epochs than in others.

In an attempt to present this suggestion visually, Forbes produced a figure which looked like nothing so much as a nineteenth-century egg-timer. The wide parts showed the era of "maximum development of generic types," meaning the maximum number of new genera, while the constriction in the middle stood for the period of "minimum production." The top and bottom of the figure were the two ends of time, spanning the first creation of living matter to the ultimate appearance of man (figure 10).

The genera introduced in these two "poles" of development exhibited a kind of representation across time. Forbes pointed out that animals such as the four-rayed corals and the Crinoidea, typical of the Silurian period in the Palaeozoic, were superseded by six-rayed corals

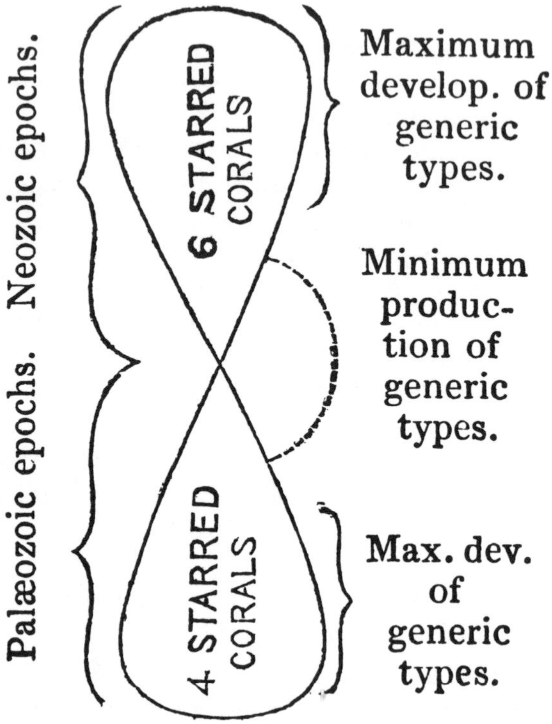

Figure 10 Forbes on the introduction of new genera over time. Fossil corals with an internal structure of four rays ("stars") were found in greater numbers in the earlier (Palaeozoic) period, while six-rayed corals reached a maximum in more recent (Neozoic) epochs. From E. Forbes, "On the manifestation of polarity," *Notices of the Proceedings of the Royal Institution of Great Britain* 1 (1854): 429.

and Echinoids in the Neozoic. The latter "replaced others of similar value and served the same purpose or played the same part in nature."[26]

The complete scheme was thought to have occurred as follows. Many genera of four-rayed corals had been created in the Silurian and were maintained through following ages by the appearance of one species after another (each of which replaced and mimicked the former, as did representative species in topographical space). Divine creation supplemented the total number of coral genera with the occasional introduction of a new genus. By the Permian and Triassic periods no new genera were being added to the coral family. From that point on, Forbes asserted, four-rayed corals were never again created and instead died out naturally, being gradually replaced by ever increasing numbers of freshly created six-rayed corals. More and more six-rayed forms were introduced until they reached a maximum of variety in the Cretaceous, Tertiary, and Modern periods.

Because this sequence of events described a maximum-minimum-maximum story, Forbes argued that the development of life proceeded in two divergent directions: "The relation between the Palaeozoic and Neozoic life-assemblages is one of development in opposite directions, in other words, of Polarity."[27]

Polarity was therefore the explanation for a progressive increase in the number of genera from a period of minimum development (the waist of the hourglass) to a maximum number of new forms in opposite

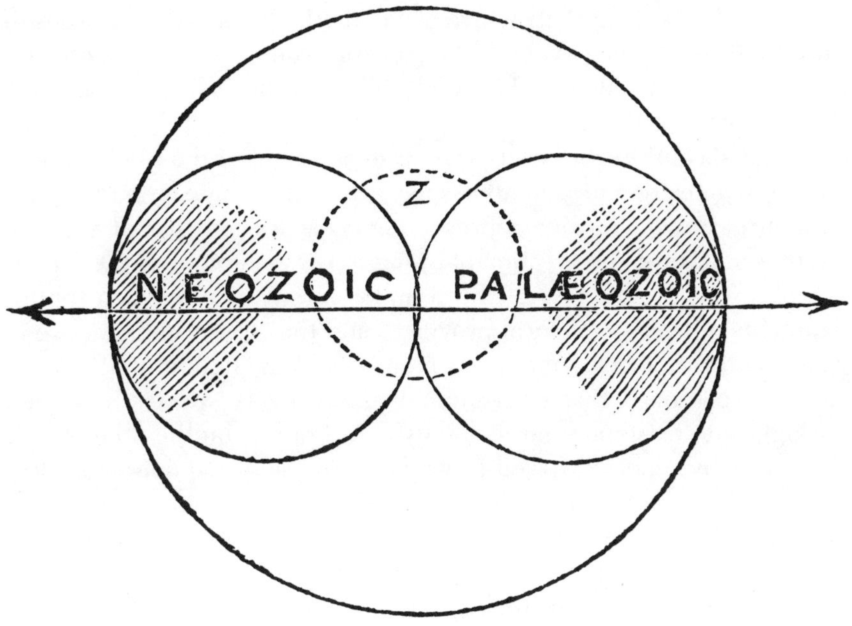

Figure 11 Forbes on the manifestation of "polarity" in the fossil record. The shaded portions of the smaller circles represented the eras when most genera were being produced, the dotted area, Z, the time of least production. The double-headed arrow signified the opposite directions in which this organic development moved. From E. Forbes, "On the manifestation of polarity," *Notices of the Proceedings of the Royal Institution of Great Britain* 1 (1854): 432.

directions at the beginning and end of geological history. If it did nothing else, Forbes's theory demonstrated that not all naturalists believed in organic history proceeding in tandem with a unidirectional or linear flow of time. His pattern of organic change stood apart from the progress of geological time, although not from changes in the environment. Forbes, as already seen, advanced the view that forms were created especially suited for the prevailing conditions and that these species changed in a regular succession according to changes in the surface of the earth. One species was replaced by another analogous species and one genus by another, precisely because the external conditions had altered and new organisms were essential. But it was the divine will that brought all living beings into existence in the first place, and, for reasons best known to himself, the Creator had increased the number of genera living on the earth at certain points in time; this plan, this scheme of things, was virtually inexplicable in terms of the geological structure of the earth. It was left to Forbes to discern order in such events. The pattern, when found, was sufficient explanation in itself. On this single point Forbes seemingly aligned himself with Louis Agassiz.

Agassiz and Forbes each in their own way used the phenomenon of geographically representative species to substantiate larger theories of nature. For the former, representative species constituted a crucial argument against the creation of a single pair of animals as the founders of a species. Because of the complex interaction between the inhabitants of every biological province, and the relationships between representative species and other members of their community, Agassiz was convinced that each individual and every species had been brought into existence simultaneously. To Forbes, on the other hand, representative species served to vindicate precisely the opposite. They went to show that although floras and faunas may have been created at the same time, their component species were called into existence only once: forms that might be expected to have a double or multiple origin were, in reality, slightly different in structure—and hence called representative species—or they were identical and their presence explicable as a result of normal migration combined with geological activity. Such an account served further ends for Forbes as a model for an all-embracing scheme of the history of life. He extended the notion of representation to cover all the interrelations of life that are now put down to evolution. In this he showed great originality, linking the spatial differentiation of species with a development through time. He believed that these phenomena were brought about by a single force in nature, and he realised that the study of one dimension would provide the answer for processes in the other. His clarity of thought on this issue rivals that of Lyell or Darwin, although Forbes's force was "representation" rather than uniformity or evolution. It is only because the system is couched in unfamiliar words and draws on transcendental metaphors that it seems strange today: to Forbes himself the principle of polarity incorporated all the hallmarks of a good scientific theory. Correlating the present and past states of living beings, it required only one pervasive force in nature that manifested itself in many different ways. It was simple and elegant, if cryptically worded. And last but not least it allowed a fundamental role for God as creator.

ANALOGY AND CLASSIFICATORY THEORY

Joseph Hooker took up the study of representative species at the point where Forbes left off. Like Forbes, he thought the phenomenon might have more to it than mere geographical replacement. Instead of using representative species as the focal point of an all-embracing theory of nature, however, he saw them as part of the ever present "species problem." Similar species of gentian may indeed replace each other in different locations in Switzerland, but how did this correlate with classificatory theory and the various morphological analogies that all species presented to some degree? What could these organisms tell him about the nature of species?

For Hooker, as for other systematists, the topic of representative species was full of complexities. The term covered manifestations of analogy and replacement in nature now thought to be unconnected with geographical races. Facts that modern biologists can separate and categorise as different phenomena were not at all separate to pre-Darwinian naturalists, thereby making a difficult subject even more complicated. Buffon, for example, wrote extensively on the ecological replacement of one species by another, citing the marsupial mole as taking the place of its placental analogue in Australia. He also referred to the more general notion that every landmass possessed local forms of cattle, sheep, and Felidae. Each of these families in his view represented those of other regions.[28] Again, Buffon had noted that some countries have different organisms fulfilling the same role in nature—as, say, the kangaroo, in the absence of native cows or horses, being the largest herbivore in Australia. One of Buffon's greatest admirers, James Cowles Prichard, furthered the analogy by describing the llama as a member of the horse genus, simply because it occupied an otherwise vacant place in the natural economy of South America.[29] Evidently there was difficulty in discriminating between ecological pairs, geographical races, the products of adaptive radiation, and those of convergent evolution.

Nor was this the only source of confusion. The word "representative" was open to many shades of meaning, particularly when used as an alternative to the highly ambiguous terms "analogue" and "analogous." In this respect it served to describe some crucial ideas in comparative anatomy. Transcendental taxonomists such as Richard Owen, William Swainson, and Louis Agassiz believed all vertebrates to "represent" some common plan, a plan reflected to various degrees in the vertebral structure so that every animal could be considered as a variation on a central theme. William Swainson developed these ideas (in his case, mostly taken from W. S. MacLeay's "Quinarian" system of classification) into a full-blown theory of representation, where each member of a discrete group exhibited a collection of qualities found individually in the other members.[30] For most naturalists of the time the issue of representative species was therefore inextricably mixed up with ideas of morphological similarities and affinities, with notions of a wider and more metaphysical nature. All these could be listed under the general heading of classificatory theory, a subject on which Hooker had a decided opinion of his own.

Hooker's peculiarly restricted upbringing—tantamount to a professional training—among the herbarium sheets of his father's ever increasing botanical collection, together with his innate sense of precision and order, had made him into an exact and uncompromising naturalist. Undisputed master of his chosen field, he was absolutely at home among the minutiae of classification tables and comparative morphology, and had gained an extensive knowledge of the floras of far-flung regions chiefly through cataloguing and arranging the specimens collected on his voyage around the world in 1839–1844.

It was a clear principle with Hooker that classification was essentially a practical matter, enabling botanists to determine exactly what plants they had before them. He felt this was best achieved by placing, as far as possible, plants with a maximum number of common characters in closely related groups. In other words, he subscribed to the natural viewpoint of French and British botanists, who had erected on the Jussieus' proposals an elaborate system of relationships, affinities, and similarities in an attempt to describe the web of relations interconnecting the plant kingdom. Hooker, always a chauvinist in botanical affairs, asserted that these points were rather more recondite than those which drew the animal kingdom together. The latter exhibited its relations through largely external features, like legs and skin, or striking anatomical bits and pieces, such as bones, guts, reproductive organs, and nerves. Botanical systematists, on the other hand, depended on the number, relative size and position of stamens and stigmas, petals, ovaries, and seed coats, and often relied on developmental or embryological factors for their identifications. All of these, he thought, necessitate subtlety and discernment in the naturalist.

To Hooker bad classification was synonymous with bad botany. And worse than both was bad comparative anatomy. So, as an earnest youth and even more earnest adult, he set about retrieving this distressing state of affairs through the example of his own meticulous work. Nor did the sense of responsibility ever leave him. Later in life he sat on a number of important scientific committees, where he was able to influence educational policies and, like his close friend Huxley, enjoy the politics and skirmishes of the government of science.

Hooker made a virtue out of his extraordinarily wide knowledge of botanical detail by claiming to be "untrammelled by any theories to combat or vindicate"[31] and by refusing to endorse anything but the most general of statements. He looked at every side of a problem in his writings. It is no surprise to find that he could not make up his mind about representative species or about the more pervasive phenomenon of degrees of representation throughout the plant kingdom. He had a clear idea of the undisputable incidence of different species of a genus appearing in geographically disparate regions; indeed, having worked on Darwin's Galapagean plants he could not fail to have noticed this particular distributional point. But when, in 1846, he was called upon to publish the results of his examination of Darwin's collection, he was at a loss to explain why so very many representative forms had come into existence.

Certainly, Hooker noted, the appearance of representative organisms was the most striking feature of the Galapagos flora: "As is the case with the fauna, many of the species, and these the most remarkable, are confined to one islet of the group, and often represented in others, by similar, but specifically very distinct congeners." He listed the species to be found in each islet and remarked that the difference between them was very odd:

> Were this peculiarity effected only by those species which may have come from the continent of South America, it would have admitted of some explanation, so capricious are the elements which regulate the interchange of species, and so uncertain in their effects even when apparently most uniform in their action. But in this case, the difference is most marked in the distribution of the species that are Galapageian only, the individuals of which are not common to every part of the archipelago, but for the greater part confined each to one solitary islet.[32]

To illustrate the point, of some 128 species peculiar to the archipelago, only 13 were found on two islands, 2 on three islands, and a single species on four. All the other plants were special to their own islet and therefore constituted representative species. Hooker could only suggest that this might conceivably be due to a lack of inter-island communication, something Darwin had pointed out in his *Journal of Researches* in 1838. Hooker thought that if all the islands were put together, within a few years they would have lost much of their individual character:

> The strife with its neighbours for position, which marks all stages of the life of any two or more contiguous plants, would terminate in a few replacing the many. . . . It must be admitted, that the first steps towards ensuring the continuance of many species in a given area, are to isolate them, and to cut off the means of migration; exactly as in a garden the plants are protected from encroachment mechanically, and the seeds of the more volatile collected betimes, to prevent a like effect being naturally brought about.[33]

And that was all he had to say on the matter. A creationist reading these words would understand them to mean that the plants of the Galapagos had been created each for their own islet. Any alteration of the present geological configuration would lead to a mixing of the flora and to strife of the kind conjured up by the elder Candolle. Yet here, through his own subtle wording, Hooker laid the basis of an alternative interpretation, one that likened representative species to the products of cultivation and hence linked this question to the wider issue of the origin and maintenance of variation. Hooker pointed to the similarities between the geographical isolation of the Galapagos archipelago and the mechanical and physiological barriers that prevent any mixture of garden stocks. If isolation kept these forms distinct, then it was possible that isolation had a part to play in the origins of such beings. Perhaps the archipelago had once been a single landmass covered by an undifferentiated flora. With the subsequent separation into several islets the plants may have formed new races through individual sporting or the fine tuning of their structures to a changed environment. Hooker seemed to favour this interpretation in explaining parallel cases of variation in, say, the Sandwich Islands,[34] and it was expressed more formally by the younger Candolle in his *Géographie botanique raisonnée* of 1855. Candolle suggested that the breakup of a large landmass into smaller units would necessarily lead to the extinction or local extermination of one or more species and to the differential preservation of others, so causing major discrepancies between island floras. Other changes arose through morphological variation.[35] He reneged on this view though, as Hooker was quick to point out in a review for the *Kew Journal of Botany* (1856), by admitting that such secondary causes of diversification were supplemented by a special creation of species.[36]

Candolle had foundered on the notorious "species problem." There were no clear-cut rules at this time for the demarcation of varieties from species, except for an intuitive sense that the former were more transitory than the latter. Varieties, and even relatively stable forms like geographical races, were plainly related to their local environments in a causal manner; should naturalists therefore reclassify representative species as mere variants, thereby acknowledging the influence of physical surroundings on organic beings? Seasoned botanists like Hooker and Candolle would recoil from anything of the sort since the organisms were plainly true species. But were these *species* formed by the same secondary causes as were mere varieties? Here they wavered. It did not help to know that the several islands of the Galapagos, for example, were virtually identical in soil, topography, and climate, and that according to their volcanic nature they must also be of fairly recent origin. Even if botanists were persuaded that the environment in question was essential, it was not certain that surroundings alone could explain the marked differences between specific forms.

It was this dilemma over the actual origin of species that led Hooker to concentrate on representative forms and the districts in which they appeared. He noted that these species often belonged to endemic genera. Writing to Darwin in 1845, he remarked that most of the representative species of the Galapagos belonged to genera also confined to that archipelago.[37] Evidently their distinctive nature was something to do with endemicity, or, rather, with the existence of some genera and species that are absolutely confined to one region of the globe.

Of course in one sense this was true, since all representative species are, strictly speaking, endemic in their own area, otherwise they could not be mutually exclusive. But Hooker enlarged the vista to discuss the problem in terms of greater taxonomic categories usually restricted to one region or another. In effect, he answered queries about representative *species* with illustrations drawn from endemic *genera*, as in, for example, a discussion of representatives in the restricted groups of cacti (endemic to the New World) and of the capuchin monkeys (limited to central America).[38] In letters to Darwin he continued to associate representative species with the attribute of belonging to a purely localised family or genus.

This emphasis on endemism led Hooker to combine relatively straightforward topographical representatives, like the various species of the Galapagos, with rather more philosophical speculations about classificatory types and typical forms. Endemism meant that certain morphological groups were special to their areas, and by implication typical of that region: koala bears were endemic to Australia, and ipso facto were a typical part of its fauna; the Japanese cherry tree typified its country of origin, as did the lion Africa. Hooker coupled this commonplace idea with his notion that representative species belonged to endemic groups, and asked himself if representatives were typical forms or if types were the same as proper representative species. The flora of the Galapagos was certainly distinctive enough to typify the archipelago: all the species were peculiar to that region, as were many of the genera; furthermore, each island was typified by its own form of local species. Typicality and representation seemed to go together in any consideration of endemic organisms.

But there was another meaning for the word "typical," which linked it not to endemic genera but to very widespread organisms which by their size and range seemed to be important and characteristic members of their group. Such dominant taxa could legitimately be called either "typical" or "representative" species.

Much of this association stemmed from the several meanings of the words "representative" and "type," a confusion by no means untangled today. A representative can be both an isolated member of the masses or an examplar. In the second sense it is a synonym for the type species of a group, on which explanations of the general structure of a family or order can be based, as the pigeon might typify the class *Aves*. This simple usage of "type" appeared regularly in the didactic writings of naturalists from Buffon to Huxley. But for the philosopher-naturalists of the first half of the nineteenth century, it also came to mean something along the lines of a model: a unifying "ideal" type for some, such as the transcendental comparative anatomists Owen and Agassiz; and a perfect living "example" for others, such as Hooker.[39] Hooker searched in vain through the genus *Cedrus* for a physical incarnation of the type to which all the species were related. The cedars of Lebanon, of the Atlas mountains, and of the Himalayas (the Deodar) were all variations on a central theme, but who could say which of them presented the model and which the derivatives? Not Hooker.[40] He confessed to Darwin: "I hardly know what is always meant by a typical form. The character of a group should be founded on the most important objects it contains in the œconomy of nature. The most important genus of a class is surely generally either the largest or the most widely diffused."[41]

Empirical evidence of size through the number of species and the extent of dispersal therefore determined for Hooker which genera should be classified as examples (or "types") of their larger groups. Returning to the problem of representative species, practical experience told him that the presence of many representative forms in a genus would give that genus rather more than the average number of species. Many mutually exclusive species also implied that the overall range of that genus was greater in extent than those of ordinary genera, as, for example, the rhododendron had many representative species and ranged widely through Asia, North America, and the Mediterranean. Hooker's special favourite, *Senecio*, presented a bewildering variety of forms, some specific and others merely geographical variations, and was spread virtually worldwide. Large numbers of representative species therefore gave the size and range that were Hooker's criteria for a type genus.

Yet this measure was in direct contradiction to the idea that genera that typified regions like the Galapagos were strictly limited in their geographical extent. Most representative species, as Hooker had shown in his work on Darwin's collections, were members of endemic genera. Representative species *ought* to belong to the most widely spread and numerous genus in a family, but were demonstrably members of purely local associations.

With the advantage of hindsight it is clear that this paradox, more than anything, prevented Hooker from coming to an unimpeachable conclusion about representative species. He knew too much and could always find an example to overturn a favoured theory, the exception that tested the rule. He examined representative species from every angle he could think of, probing deeply into the philosophy of classificatory theory and his knowledge of geographical distribution, and failed to answer the questions he had set himself. Hooker did not know what to think about representative species, their origins, or how to draw them into the general scheme of nature.

The solution to Hooker's problem was soon provided by his friend Charles Darwin. Indeed it was Darwin's virtuoso display on representative species that won Hooker over to the theory of evolution.[42] Darwin showed Hooker how representative species could be thought of as members of genera once widely spread but now restricted in area. By supposing that an ancestral "parent" species had originally been far-flung, reaching out over all the lands now colonised by a series of representatives, Darwin explained how a species could act as a "type" for a whole group. If and when this species split up (through the action of competition and natural selection), it would form smaller and mutually exclusive subspecies. Some of these might die out, leaving only a cluster of close allies—each "representing" the next—contained within a genus that was native to one particular location on the globe. Representative species were both widespread and restricted in their ranges, but never at the same time.

Hooker had not visualised the various phenomena of representation in this peculiarly historical way; he could see no chronological connections between a large and well-dispersed species and the constellation of similar, but mutually exclusive forms that were called representatives. Before Darwin introduced him to evolutionary theory, Hooker was at a loss to link together these apparently opposing qualities of wide range and endemicity. Darwin gave him the concept of time, the idea of changing states and sequences that enabled one kind of species to change into many smaller, interrelated forms. Representative species, Darwin thought, could be explained only by reference to history.

This illustrates some of the differences between Hooker's approach to species and that of Forbes and Agassiz. Hooker, as a systematist, was primarily concerned with explaining what he saw in terms of current classification theory; his inability to account for representative species rested not on any personal inadequacy or lack of attention to detail, but on the contradictions and subtle shades of meaning that were becoming increasingly apparent in pre-Darwinian systematics. The *origin* of representative species had to be ignored in the face of more perplexing issues in theoretical taxonomy, such as the need to distinguish species from varieties and to provide consistent guides to the many more intermediary forms of life. By contrast, it was the origin of representative species that fascinated Agassiz, as an example of how other species were created and as vindication for his own highly individualistic theories of zoogeography. Yet Agassiz, like Hooker, did not connect the moment of origin for representative species with the present by supposing a process of gradual dispersal and change in form, as Darwin was to do. Forbes, on the other hand, did capture something of this concept of process and change. He noted a gradient of resemblances for species and for floras and faunas (without suggesting any physical links between similar forms), and he applied ideas derived from topographical studies to the fossil record, thereby producing an explanation for life in both dimensions that depended to a large degree on sequence and series. But Forbes believed in the fixity of species. He did not associate the origin of representative species with the process of differentiation that could be seen in the formation of geographical variants. Only Hooker, of these three, seemed to hold the idea that representative forms were part and parcel of the species problem, intimately linked to the production of varieties and garden cultivars. Yet Hooker's attention was diverted elsewhere, to pragmatic questions of types and typicality.

How easy it was for naturalists of the second quarter of the nineteenth century to talk past each other even though they were all investigating the same phenomenon. Plainly any concept that could resolve this multilayered issue, and at the same time unite the explanations already offered, would be most welcome. The answer was found in the idea of organic change.

SEVEN

Organic Change

During the second quarter of the nineteenth century geographical distribution had engaged the attention of both natural historians and geologists, resulting in studies that can loosely be divided into those that were topographical in nature and those that were chronological. But by the 1840s a new breed of distributional naturalist had also emerged. Raised on Lyellian geology and anxious to discover the historical processes that lay behind the geographical arrangement of life, men such as Edward Forbes attempted to fuse both dimensions into a single story. Forbes, for example, presented a system that fully integrated present-day distribution patterns with the fossil record, so making a crucial juxtaposition between distribution in space and in time and establishing a methodological principle that was basic to subsequent studies of animal and plant geography. Although Forbes was a dyed-in-the-wool creationist, he was nevertheless one of the first naturalists to understand biogeography in the larger and more modern sense of a union between patterns *and* processes.

The Lyellian club was not large: besides Forbes, Charles Lyell himself, and a slightly bemused Joseph Hooker, the only other members who expressed interest in this kind of integrated investigation were Alfred Russel Wallace and Charles Darwin. The last two were not content, however, to think of distribution solely as a sequence of changes over time and space. They added a third element to Lyellian distribution theory, namely the idea of organic change—that species themselves could alter in structure and that taxonomic groups had successively evolved, diverged, and passed away in an unbroken stream of related beings. In retrospect, Wallace and Darwin gave natural historians the third and final piece of the biogeographical jigsaw. With the advent of this evolutionary theory, biogeography developed into a subject concerned not only with changes in past and present distribution patterns, but also with modifications in the individual organisms themselves. Indeed, they supplied the foundation stones on which all modern biogeographical endeavour is based.

Alfred Wallace (1823–1913) and Charles Darwin (1809–1882) were obviously seminal to the future sophistication of biogeography as a scientific field. Without them no history of the subject could be written. But instead of asking what Wallace's and Darwin's theories did for animal and plant geography, it is perhaps more relevant to ask what the topic did for them. Why were they interested in this particular branch of natural history? How did they interpret the data of distribution, and how much did their interpretations contribute to the inception and development of evolutionary theory? More specifically, it is important to know whether Wallace and Darwin approached animal and plant distribution in an identical fashion, as is usually assumed, or whether they differed in some significant respect. The answer, it seems, is that they did indeed differ in the selection of problems to be solved, in the kind of explanations offered, and in the deeper philosophical background to their studies. Even the first formal presentations of the theory of evolution were different in their treatments of geographical distribution, thereby providing a clue to at least some of the events surrounding the simultaneous discovery of natural selection. Wallace's and Darwin's distributional studies provide a fascinating backdrop for the reexamination of the emergence of evolutionary theory.

WALLACE AND GEOGRAPHICAL DISTRIBUTION

Alfred Russel Wallace's simple family backgroud and relatively humble occupations form an appealing counterpoise to the startling acuity and vision of his mind. The third son of an unambitious and poverty-stricken lawyer, he had no formal schooling beyond the age of fourteen and led a peripatetic existence as assistant to his surveyor brother before settling temporarily as a master in the Collegiate School of Leicester in 1844. As a youth he read voraciously, having access to a wide variety of classics, travel books, and radical literature mostly of the Owenite school, all of which he found in the libraries of local Mechanics' Institutes. He was inspired by reformers such as Robert Owen—whom he went to hear in Leicester—and George Combe. His intellectual views were consequently moulded in a distinctly radical and "improving" environment, where an optimistic faith in progress through the operation of beneficent natural laws prevailed.[1] These views, acquired at an early age, were to remain with Wallace for the rest of his life; they were reflected not only in the theory of natural selection and his work on the future evolution of man, but also in a belief in phrenology, spiritualism, and the afterlife. He also published idealistic literature of his own on the land question and socialism.

Yet sociopolitical tracts were not the only works to interest the young Wallace. At Leicester he had read Alexander von Humboldt's *Personal Narrative*, which filled him with a burning desire to travel and experience the wonders of the tropics. Darwin's *Journal of Researches* added fuel to the fire. Other books on the self-imposed reading list included Thomas Malthus's *Essay on the Principle of Population*, then in its sixth edition (1826), and Robert Chambers's anonymously published *Vestiges of the Natural History of Creation* (1844). Both of these were to have a profound effect on Wallace's later thinking.

Although he was still showing little interest in geographical distribution at the close of the 1830s, Wallace had become fully immersed in the subject by 1842. In that year he read William Swainson's *Treatise on the Geography and Classification of Animals* (1835). As was proper for an elementary publication in a *Cyclopaedia* series, Swainson set out the basics in a clear and attractive manner. Notably, in view of Wallace's later concerns, he indicated the variety of theories held by contemporary scholars and pointed to their inadequacies. Passing from Linnaeus's dictum of one source for all species (apparently "controverted by the facts"), Swainson described J. C. Prichard's alternative of a postdiluvial creation, each animal in its own region.[2] This too was unsatisfactory to Swainson's way of thinking, even with Prichard's half-hearted qualification that the flood may have been a local event. Buffon's theory that organisms were specially adapted to their own countries by variation was thought to be negated by the fact that species could readily be introduced to other lands without any change in form. Swainson concluded that the phenomenon of geographical distribution was ultimately impossible to solve. It did not depend simply on secondary causes like adaptation to local conditions, but rather on a single primary cause which had placed some animals in one hemisphere and not in another. God, he asserted, created an original breeding couple for each species; from separate points every species had multiplied and diffused to cover the area it now occupied. Beyond this explanation the philosopher could not hope to go. The original creation, and the reason why one organism went with a particular region, was an obscure mystery: "The primary causes which have led to different regions of the earth being peopled by different races of animals, and the laws by which their dispersion is regulated, must be forever hid from human research," Swainson wrote.[3]

To a restless and inquiring thinker like Wallace, this judgment must have presented an immediate challenge; Swainson was virtually challenging him to take up geographical distribution and the study of creation. Unhindered by any religious preconception, disinclined to leave everything to God, Wallace had long ago abandoned orthodox Christianity in favour of the "natural laws" of Paine, Combe, and Owen. He disparaged Swainson and the others for "attempting to reconcile science to scripture." Had a creator really placed animals in regions for which, in Swainson's words, they are by their nature peculiarly adapted? "No," asserted Wallace in a vigorous marginal note.[4]

But it was not until three years later, in 1845, that Wallace came across another book dealing with the organic world that echoed his own doubts about an interventionist creator. Robert Chambers's *Vestiges of the Natural History of Creation* proposed an evolution of organisms through entirely natural processes of reproduction and variation, set in motion at the dawn of time and left to operate independently of an "August Being." Surely it was ridiculous, claimed Chambers, to expect a Deity "to interfere personally and specially on every occasion when a new shell-fish or reptile was to be ushered into existence?"[5] It was more seemly (though few of Chambers's readers agreed) to draw a parallel between the perpetual, self-regulating world of astronomy and that of the history of life; to assert that God had devised all-embracing natural laws for the organic realm as well as for the physical arena. These laws acted on the raw stuff of living matter to produce a succession of forms from mere "globules" and simple cells, up to the most organised of beings. Men, too, evolved from lower forms of life and were still evolving toward a higher and more godly state.

Chambers's world was patently governed by universal and beneficent natural laws. He built the *Vestiges* around an ideology that Wallace had already encountered and willingly absorbed. Then still only an amateur in the field of natural history, Wallace had no way of assessing the validity of Chambers's facts, nor of evaluating his arguments, yet he was captivated by the thesis so many of his contemporaries despised. He took up the idea of progressive development. "I well remember the excitement caused by the publication of the *Ves-*

tiges," wrote Wallace at the close of his life, "and the eagerness and delight with which I read it. Although I saw that it really offered no explanation of the process of change of species, yet the view that change was effected, not through any unimaginable process, but through the known laws of reproduction, commended itself to me as perfectly satisfactory, and as affording the first step towards a more complete and explanatory theory."[6]

Chambers's argument was mostly derived from an extended description of the geological formations of the earth and the fossils found therein. In tune with many geological theorists of the age, he presented these facts as a history of a gradual development over time, each epoch possessing animals and plants suited to the immediate environment of a progressively cooling earth. The organic world had therefore advanced in organisation and complexity over time, more or less in linear fashion, as the earth had approached the physical conditions of today. Chambers departed from this run-of-the-mill approach, however, when he stipulated that the species of each succeeding formation were descendents of those which preceded and that organisms had evolved one from another. About the actual mechanism—as Wallace duly noted—Chambers was more than a little hazy, saying only that variation usually accompanied reproduction. Such a "law of development" would suffice, he thought, to make one species appear from another. An unfortunate example was that a goose may be supposed to have given to its progeny the body of a rat and thence produced a duck-billed platypus.[7]

In the first edition of the *Vestiges* in 1844, Chambers substantiated his transmutation argument with evidence taken from classificatory theory and, in passing, from geographical distribution. His account of the latter was brief but incisive. The book pointed out that each biological region of the earth possessed different sorts of life that must, on the theory of progressive development, have evolved *in situ*. Local conditions, suggested Chambers, stimulated variation with the result that species became suited to their environment. Should conditions be eminently favourable for particular forms of life, these would respond by speciating in greater numbers, "pressing in" whenever they could find room and refreshment. Chambers did not mean here that regions became fuller, so to speak, but that in certain times and places a proliferation occurred of new species and genera belonging to the same family or order. His own example was that "the limestone-forming seas produced an abundance of corals, crinoidea, and shell-fish."[8] This unusual view, which was to prove deeply significant for Wallace, harked back to Edward Forbes's comments on the relative numbers of genera over time. Forbes, too, felt that some eras had experienced a greater production of specific and generic forms than had others. Chambers evidently thought that when conditions were right the lower taxa multiplied, fleshing out the skeleton, as it were, of evolutionary change.

Wallace was impressed. He shared this approval with his friend, H. W. Bates, the entomologist, saying,

> I do not consider it as a hasty generalisation, but rather as an ingenious hypothesis strongly supported by some striking facts and analogies but which remains to be proved by more facts and the additional light which future researches may throw upon the subject. It at all events furnishes a subject for every observer of nature to turn his attention to; every fact he observes must make either for or against it, and it thus furnishes both an incitement to the collection of facts and an object to which to apply them when collected.[9]

It is clear that, with these thoughts uppermost, Wallace did not travel to the Amazonian basin simply for adventure or to profit from the sale of specimens. He also went to gather information about the formation of species. Whether he had the germ of a theory of natural selection is debatable, but it seems that he was already convinced of the reality of evolution. Like Darwin, he believed in transmutation for some years before he discovered a suitable mechanism. From that time on, Wallace was always in pursuit of an "explanation of the process of change of species."

But Wallace's voyage with Bates to the river Amazon, from 1848 to 1852, did not provoke any definite thoughts on the evolution of species, although it did provide ample resources for his first serious work in geographical distribution. Wallace noted the distribution of butterflies and birds as determined by the presence or absence of geographical barriers, and in some short contributions to British periodicals as well as his *Travels on the Amazon and Rio Negro* (1853) he related many of these biological patterns to the pronounced influence of river systems. Three species of the genus *Psophia* illustrated his argument: the green-winged trumpeter bird lived only on the south bank of the Amazon and east of the river Madeira as far as the forests of Para; the white-winged form lived exclusively west of the Madeira; while north of the Amazon the common trumpeter bird took over.[10]

Privately, Wallace may well have understood this as the result of evolution after geographical isolation, but (for one reason or another) he failed to present it as such in his published record. At that time he had no mechanism for evolutionary change; even if he had, the *Travels* was hardly an appropriate vehicle for the exposition of revisionist natural history. Sticking closely to a narrative format, Wallace's book was not overly theoretical even though it was supposed to be based on those two classics of travelling natural history, Humboldt's *Personal Narrative* and Darwin's *Journal of Researches*. Nor was it rigorously scientific. In fact Darwin felt this semipopular style was a mistake. He complained that Wallace supplied little scientific information, "hardly facts enough" for the philosophical reader.[11]

Darwin appears to have spoken for the majority of the scientific community in his judgement of Wallace's work, for the younger man's book won only superficial praise. Feeling that he had been spurned by the scientific elite of Britain, Wallace decided after fifteen months in England to set off once more into the wild. This time, however, he selected the Malay archipelago, the decision being influenced by the knowledge that European museums would pay well for specimens from this little-known part of the world. He was by now supporting, as far as possible, the remaining members of his family and the economic aspects of the trip were not unimportant. Furthermore, the area also promised rewards of an intellectual kind. In the *Vestiges* Chambers had drawn attention to Indonesia as a probable site for the evolution of man: there, many primitive tribes occupied the same localities as the Orang-utan, the implication in evolutionary terms being that one set of living beings had given rise to the other. The ethnology of the area therefore beckoned. Additionally, although it was not entirely unknown, the animal and bird life of the archipelago had never been examined methodically. A scientific reputation might well be founded on a suitably full and accurate account of the Malaysian fauna. But in the final analysis Wallace no doubt recognised that he had developed a taste for the riches of the tropics, and for the solitary life of a travelling naturalist. It was this, more than anything, which drew him back to the rigours of another exploratory voyage.

So, with the financial aid of the Royal Geographical Society, he began in 1854 a trip which lasted until 1862, during which he composed the two papers that so dramatically influenced the course of his biological career. In these he put forward his own account of "progression" and "development."

REFUTING THE PRINCIPLE OF POLARITY

Wallace first aired his views on the origin of species after the publication of Edward Forbes's theory of polarity in 1854. "It was the promulgation of Forbes' theory which led me to write and publish for I was annoyed to see such an ideal absurdity put forth when such a simple hypothesis will explain all the facts," he wrote.[12] A paper was consequently composed during his time at Sarawak, in February 1855, forwarded to England, and published in the *Annals and Magazine of Natural History* in September the same year.[13] Forbes had claimed that species and genera occupied discrete areas in topographical space and geological time, forming groups that replaced and mimicked each other over the globe or through the fossil record. On the contrary, asserted Wallace, these sequences were not due to an abstract metaphysical "representation" of one set of beings by the next, but were instead due to a real relationship of affinity: "every species has come into existence coincident in both time and space with a preexisting closely allied species."[14]

This "simple hypothesis" was derived from the *Vestiges*' law of development, with a dash of Charles Lyell's *Principles of Geology*. In his Sarawak paper Wallace briefly portrayed the history of the earth as one of constant and gradual change, prior to stating that the "present distribution of life upon the earth must be the result of all the previous changes." Like Forbes—and like Lyell—he recognised that the world had been an active theatre and that the cast and scenery had altered over a period of many years. The pattern of today was only one in a changing sequence of organic and geological fluxes. Assuming that physical barriers were the prime factor in limiting the spread of organisms, he pointed out that an area isolated from other faunas for a long time ought to possess distinctive features of its own. The most important fact for Wallace was that the families and orders found in such strictly defined regions possessed many species and genera. There had been an explosion, as it were, of closely allied forms all organised on the "type of pre-existing ones."[15] In modern terms, he supposed that isolation was followed by adaptive radiation and speciation, a situation perfectly illustrated by the bird life of the Galapagos. Indeed, Wallace went on in his paper to explain how the Galapagos archipelago had been populated by emigrés from the South American continent (his example being taken from Darwin's *Journal of Researches*), which had subsequently diversified and colonised the islands.

Wallace undoubtedly got his idea of a proliferation of lower taxa from the author of the *Vestiges*. Chambers had made much of it, describing how life seemed to "press in" whenever conditions were particularly suitable, an activity reflected not in the creation of many new orders or families but in the multiplication of lesser taxonomic categories. Naturally, if all these belonged to one family they would be constructed on the same general plan, or, as Wallace called it, according to an ancestral "antitype." Wallace considered Chambers's fossil explosions as part of a process that could also be seen in topographical space and believed that circumscribed geographical regions ought to possess great numbers of closely related forms: the Amazon basin, for example, clearly presented three species of trumpeter bird, showing some kind of increase in the number of minor taxa. Species evidently multiplied when the conditions were right, either in time or space.

Wallace saw this proliferation as a diversification of what was already there, an approach that led him into conflict with Edward Forbes. Forbes had thought that certain geological periods were marked by the introduction of *new* genera, forms entirely different from those which came before and after. Any similarity between the organisms of such groups was, for Forbes, one of metaphysical resemblance only. Divine ideas do not evolve one from another. For a pragmatist like Wallace, unfamiliar with or unimpressed by the philosophy of ideal types, the similarities were nothing else but real. The resemblances between the many closely allied species of a particular topographic locality, and the similarities between those of the fossil record described by Chambers (in turn echoing Lyell), led Wallace to dictate his now famous rule: every species has come into existence "coincident both in time and space with a pre-existing closely allied species."

Nowhere in the 1855 paper from Sarawak does Wallace pretend that he is offering a transmutationist view, but it seems clear that this is what he had in mind. Many years later he recalled that the piece "clearly pointed to some kind of evolution. It suggested the *when* and the *where* of its occurrence, and that it could only be through natural generation, as was also suggested in the *Vestiges*."[16] Indeed it would be hard to interpret the paper in any other sense. Wallace described the significance of his new law for amplifying our understanding of the classificatory tree and fossil record, and furthermore explained the existence of rudimentary organs as shadowy leftovers of a process of organic change. His work trembled on the brink of outright transmutationism.

Certainly, Charles Lyell was taken aback by the mutationist overtones of this theory which so plainly intermeshed with his own views on geological history. Lyell had gone to great lengths to refute Lamarckian doctrines in the *Principles*, using among other things a sophisticated account of the countless shifts in geological structure to make the transmutationist position look absurd. Here, however, was a young man employing uniformitarian arguments to validate a new version of the same thing. Thoughtfully, on 28 November 1855, Lyell opened a notebook to examine the "Species Question." Wallace's name appeared at the head of the first page of text.[17] For six months Lyell argued with himself about Wallace's paper, focussing on the mutability of species and the central assertion that species come into existence "coincident" with others closely related to them. Lyell disagreed with the implications of Wallace's arguments and firmly told himself that "there are innumerable reasons connected with the past and future as well as the present which will cause the new species to resemble those which exist or which lately existed."[18] His own theory of creation stipulated that species presented close resemblances because they occupied roughly the same "place" in nature: geological changes were such that the same sort of conditions were repeated in both time and space, although they were never *precisely* the same due to infinitesimal variations in physical and organic factors; the same species could never be created twice for this very reason. But as geological change proceeded he supposed conditions to alter to some small degree, necessitating the creation of another slightly different species to take the place of the former.[19] With this emphasis on special creation, Lyell's views were diametrically opposed to those of Wallace.

Lyell spent a restless winter grappling with the implications of Wallace's theory. Eventually he took the problem to Darwin, whom he knew to hold considered ideas about the species question. A meeting between them on 16 April 1856 resulted in Darwin telling Lyell of his own answer to this troubling issue—the theory of evolution by natural selection.[20] Immediately, Lyell urged Darwin to publish without delay. He may even have warned Darwin of Wallace's interest in this subject, pointing out how similar the two approaches appeared to be.

As it turned out, Darwin had already read Wallace's 1855 paper, which he had dismissed as merely another creationist tract. Wallace, however, was patently on the opposite tack, as is made plain by his own notebooks of the period 1855 to 1858, in which he fully espoused the development doctrine and noted how geographical distribution and geology combined with it to form a series of changing populations over time. He lacked only a hypothesis for the actual mode of evolutionary change.

This he concocted, so the famous story goes, while lying in a fever on the island of Ternate in the Moluccas. In January or February 1858, brooding on the problem of species, he remembered the work of T. R. Malthus and with a great leap of the imagination applied Malthus's principle of the development of human societies to the animal world. Wallace envisaged that extinction on the one hand, and selection on the other, would lead to a gradual improvement of the race and to a "tendency of varieties to depart indefinitely from the original type."

SPECIES AND VARIETIES

The intellectual background to Wallace's imaginative leap demands renewed attention from historians. Like Darwin, of course, Wallace had been on the verge of this breakthrough for some time without realising it. Even more interesting is his use of Malthus's *Essay on Population*, the same book that gave Darwin the idea of natural selection; for Wallace, as for Darwin, Malthusian arguments were the agent by which various lines of thought were brought together and converted into something remarkably new. But it is not generally known how Malthus's writings interlocked with Wallace's hypotheses or, indeed, what Malthus may have meant to him at that critical feverish moment. As in so many other areas, the spectre of Darwin has overshadowed and obscured important aspects of Wallace's scientific accomplishments.

A series of interrelated problems was troubling Wallace during his years in Malaysia. His Sarawak paper (1855) on "the law which has regulated the introduction of new species" had set him thinking more and more about the problem of organic change, and particularly about the proliferation of species. Using Chambers's description of "explosions" of lower taxonomic groups at certain times in the fossil record, Wallace had provided ample evidence that the same phenomena occurred in the natural world of the present day; he had shown that over a well-defined geographical area (such as the Amazonian basin) where the conditions are favourable there will be many closely related species in a family (such as the Lepidoptera). These species, he implied, had evolved or developed from some other single ancestor. An aboriginal form may have covered the whole basin only to be separated at a later date by various waterways. Such isolation would automatically preserve any changes that might arise in the structure of each secondary population, leading to the formation of closely allied or "representative" species.

But Wallace was not content with Chambers's ideas about the way changes in organic form actually appeared. Chambers's law of development, where one species literally gave birth to another, to him seemed crude and inadequate. What intrigued Wallace was the relative number of individuals in different varieties and species. To this issue he returned again and again, finding that apparently dissimilar avenues of thought always led back to the same central concern. It was evidently the nub of his theoretical dilemma.

In one sense, Wallace's interest in relative numbers was generated by practical commitments. As a practising naturalist, he was constantly faced with difficult classificatory decisions: organisms had to be allocated to the appropriate taxonomic category; hierarchies of distinction had to be observed; consistency had to be maintained. Most important, species and varieties needed to be clearly distinguished and described. Wallace took this last point very seriously indeed. Anxious to make a good career out of natural history, he was only too aware of the need to provide acceptable systematic descriptions: his efforts would be of no use to Wallace if his species and genera were discredited by professionals at home or if his criteria were superseded by those of armchair naturalists. Furthermore, his income and the continued well-being of his family depended on the successful sale of specimens, specimens whose value increased with rarity or scientific peculiarity yet which were virtually worthless without proper attribution. For these reasons Wallace could be relied on to classify his species with greater than average care and attention.

Like countless naturalists before him, Wallace found that in some cases it was exceptionally difficult to distinguish species from their varieties. Some varieties were so distinct in their structure, so extended in their geography and consistent in their reproduction that he would certainly have called them species if he had not known of the existence of a parent form. Alternatively, some organisms varied only slightly from the original mould, making Wallace wonder if they ought to be considered varieties at all. Either way, the dividing line between species and varieties was indistinct where Wallace would have wished it to be certain. He began to ask himself whether such a dividing line actually did exist.

Wrestling with this problem of categorisation in his diaries and notebooks, Wallace felt the need for expert advice. In October 1856 he wrote to Darwin about the nature of species, Darwin at that time being virtually unknown to Wallace except as the author of important books on geology and zoology. Wallace asked for Darwin's opinion on the practical distinction between varieties and species and added a series of related questions on physical changes brought about in domestic animals and on the effects of climate. The younger man plainly hoped that Darwin had read his paper in the *Annals* and would now enter into scientific correspondence with him about the species problem.[21]

Darwin's reply was noncommittal, even evasive. He told Wallace, on 1 May 1857, that he had been working on that very problem for nearly twenty years, on "the question how and in what way do species and varieties differ from each other." Rather than elaborate on the subject, however, he insisted that "it really is *impossible* to explain my views (in the compass of a letter) on the causes and means of variation in a state of nature."[22] He had read Wallace's paper and agreed to "the truth of almost every word." He too had deliberated on the same problems for many years. With his particular interests thus firmly established, Darwin retreated into polite scientific questioning. He had no intention of disclosing more than was absolutely necessary; as far as he was concerned the topic was discreetly closed.

If Darwin's letter was meant as a warning to Wallace, the younger man did not notice. Flattered that Darwin took an interest in his work, he wrote to H. W. Bates saying that Darwin was preparing a great work on species and varieties and that it would undoubtedly provide a large number of facts to support Wallace's personal theories. The important thing for Bates to note was that Darwin "may save me [Wallace] the trouble of writing the second part of my hypothesis."[23] In other words, Wallace believed Darwin was about to demonstrate a physical continuity between species and varieties, the very problem he had been struggling with during recent months and to which he had as yet found no solution. It was much more convenient, he implied to Bates, if an important scientist such as Darwin established the point; when Wallace came to write he would be able to take this taxonomic theory as an accepted fact.

And yet Wallace had by now acquired his own views on the continuity between species and varieties. By the middle of 1857 he was convinced that there was no theoretical consistency in these taxonomic categories. The practical difficulties encountered in the classification of lower taxa were further enlarged by logical, abstract discrepancies arising out of systematic theory itself. The proper definition of a variety was at odds with the current definition of a species, the problem being exacerbated by so-called intermediate groups such as geographical races and representative species.

Wallace set out his views in a "Note on the Theory of Permanent and Geographical Varieties," published over a year later in the *Zoologist* for 1858.[24] A species, he declared, is supposed to have permanent taxonomic characters, whereas a variety (as its name suggested) ought to be more variable. What, then, was the naturalist to make of varieties that were obviously permanent in form, such as geographical races of birds or beetles? If it was acknowledged that the characteristics of such groups were permanently fixed, it would mean that these and all other varieties were in essence little species—that they differed from species only in the number of characters that were permanent or in their degree of distinctness. There could be no fundamental difference in kind. Consequently, for Wallace at least, the theory of special creation was called into doubt. Special creation required that species possessed some quality unknown in varieties; if no separate feature can be found, that fact is one of the strongest arguments against the independent creation of species. "Why," asked Wallace, "should a special act of creation be required to call into existence an organism differing only in degree from another which has been produced by existing laws?"[25]

To Wallace, of course, it was clear there were many permanent varieties in the world: butterflies, for example, presented an infinite number of geographical races, each strictly fixed in character and readily identified by stable diagnostic features. Much of Wallace's daily experience testified to the existence and stability of geographic races in a variety of classes. To him, such forms controverted the doctrine of creation. They revealed logical inconsistencies in a taxonomy based on divine intervention and proved there was no real difference between species and varieties. This point, more than any other, explains the emphasis Wallace laid on geographical races and representative species at that time in his life. They were crucial because they supported the first major step in his search for the mechanism of evolution. Through geographical variation he was convinced that species were essentially the same as varieties.

Having established that point to his own satisfaction, Wallace was left with the problem of deciding what, then, were the factors that distinguished the lower taxa from each other. Since he discounted qualitative differences, all that remained were quantitative ones. He was forced to the conclusion that numbers alone could determine systematic status. The smaller population was inevitably designated a variety of the larger, even though its distinguishing characters might equal or surpass those of the more numerous association.

WALLACE'S THEORY OF EVOLUTION

It seems that by the end of 1857 Wallace was preeminently concerned with the numerical or statistical side of his researches. Everything appeared to depend on the actual quantity of individuals found in this group or that. Moreover, underlying and reinforcing this numerical statement about individuals was Wallace's belief that larger groups like species can increase in number (relative to genera) when conditions appear favourable, creating a flurry of twigs on the tree of life. Numbers of one sort or another seemed fundamental to understanding the natural world. Luckily for the taxonomist, nature appeared to be stable or, more properly, in a state of equilibrium. If, however, there was to be change in the history of life, there would also have to be change in the relative numbers of individuals, as might occur, for example, in the case of a variety overtaking and eliminating its own species or of one species diversifying into several others. Wallace must have turned to consider the factors that regulate the abundance of organisms.

Here, Malthus was an obvious source. Although exclusively concerned with human populations, Malthus's fame rested on an exhaustive discussion of the ways in which one set of individuals increased at the expense of another. He dealt with the relative numbers of the poor and the rich; with their respective rates of increase; and with the social chaos he believed would result if the birth rate of the lowest classes was allowed to flourish unchecked. Malthus on population was a superlative study of relative numbers. It is hardly surprising, given Wallace's interests at this critical juncture, that he should think of such a book. Nor should it go unmentioned that Wallace had easy access to Malthus's ideas through the pages of his travelling library. There, in the second edition of Darwin's *Journal of Researches* (1845), Wallace could find a resumé of the Malthusian thesis directly applied to the problem of animal population sizes, their apparent equilibrium, and their ceaseless fluctuations. In a little-known revision of the first edition Darwin had added paragraphs on the conditions of existence and the causes of extinction, pointing out how these determined the relative abundance of given species.[26] It would be ironic indeed if Wallace had turned to Darwin's remarks in search of information on varieties and numbers of individuals and rediscovered the laws of checks and balances that would lead him on to the concept of natural selection. But this is mere conjecture.

More to the point, Malthus provided Wallace with the key to "the exact process of the change [from species to species] and the causes which led to it." The eighteenth-century author asserted that food and space are strictly limited and that growth of any population must eventually bring the number of inhabitants to its ultimate limit. The population cannot grow forever, and something must bring the growth to an end. Malthus proposed the dismal theorem that equilibrium was brought about by misery, starvation, disease, and poverty—positive checks on the poorest classes—and by moral restraint in the higher reaches of society. Such "checks" ensured that only a fraction of the total number of infants born survived to a natural old age.

For Wallace the theory meant that the numbers of animals and plants depended on the food resources of their native habitats. Should food supplies increase, individuals would increase; decrease, and there would be competition for whatever remained available.

And here was Wallace's true inspiration. He decided that those individuals more successful in obtaining food or better adapted to surmount natural obstacles would be the ones to survive. Birds with stronger wings than the rest would, in times of hardship, be able to fly further in search of grain. Wallace believed this argument to explain the unparalleled success of the passenger pigeon:

> Why is this bird so extraordinarily abundant, while others producing two or three times as many young are much less plentiful? The explanation is not difficult. The food most congenial to this species, and on which it thrives best, is abundantly distributed over a very extensive region, offering such differences of soil and climate, that in one part or another of the area the supply never fails. The bird is capable of a very rapid and long-continued flight, so that it can pass without fatigue over the whole of the district it inhabits, and as soon as the supply of food begins to fail in one place is able to discover a fresh feeding-ground. This example

> strikingly shows us that procuring a constant supply of wholesome food is almost the sole condition requisite for ensuring the rapid increase of a given species.[27]

The same reasoning explained why sea birds survived in great numbers: not because they were more fertile or prolific than other birds, but because their food supply never failed. There was for all animals a delicate interplay between the size of the population and its physical resources. Normally, wrote Wallace, species were in a state of equilibrium, having long ago reached the maximum number of beings that their area could support. Reproductive rates being what they were, enormous numbers of organisms must die annually. Taking from Malthus the idea of positive checks, Wallace supposed that it was the weakest animals—"the very young, the aged and the diseased"—that died in order to maintain the status quo. The combined effect of such high production and forcible destruction was a stable population. There could be no alteration in the numbers without some change in the conditions of existence, some shift in the geological or geographical environment that modified the food supply.

Wallace applied this argument to entire taxonomic units, as well as to individuals. If any species, he suggested, produced a variety that had slightly increased powers of preserving its existence, the variety must inevitably in time acquire a superiority of numbers. In effect, the variety would soon be competing with its parent species. If something altered in the physical conditions of the district, it was obvious to Wallace that the inferior group—in this case the species—would gradually diminish in numbers and might also become extinct. "The superior variety would then alone remain, and on a return to favourable circumstances would rapidly increase in numbers and occupy the place of the extinct species."[28]

Here, then, Wallace had his mechanism for evolutionary change. He understood evolution to take place through the successive replacement of species by varieties, the latter coming to be called species as they increased in numerical size. It was done through the improvement of one group at the expense of another and was regulated by ever vigilant Malthusian checks. Well-adapted varieties gained more and more individuals until they paralleled or exceeded the numbers contained in the parent species; geographical or other changes in the physical environment forced the parent to compete with its more plentiful offspring; searching unsuccessfully for food, the parent gradually declined in numbers and was eventually destroyed. The original variety increased to fill the gap and became a species itself. The process could be repeated *ad infinitum*, resulting in a gradual shift in organic structure.

Wallace's theory was clearly based on a statistical appreciation of natural populations. He saw that varieties and species were only distinguishable according to relative abundance and that if these numbers were reversed a variety would become a species. The idea of a Malthusian selection of the strong and healthy animal—of the most perfectly adapted form—gave him a striking explanation of the way in which such reversals might take place. Unsuccessful organisms were swept away by more successful offspring.

It is important to note, however, that Wallace did not visualise selection as a process where individual organisms competed with each other directly; he did not possess a Darwinian understanding of events, where living beings vied among themselves and one by one became better adapted to surrounding conditions. Wallace's selection did not take place at the level of the individual per se, but at the level of varieties and species. He believed evolutionary change was brought about by the replacement of one group by another, by one taxonomic unit taking the place of the next. His varieties (distinct from mere individual variants) were thought to be more suited to the immediate environment than the parent; this was why they increased in number and eventually superseded the original association; when the new species produced further varieties, it too would succumb in its turn. The process was one of increasing adaptation and adjustment of an entire *taxon* to local environmental factors.

For this reason, natural species simply could not regress back to some aboriginal state. Every step they gained they had to hold, otherwise the whole group would soon become extinct. There was no analogy, therefore, between changes taking place in nature and those seen under domestication. Varieties produced by purposeful breeding were adapted, as it were, to artificial surroundings; they could not survive in the real world unless they returned to some basic, natural prototype. Domestic breeders were incapable of eliminating this natural prototype, nor could they completely and permanently divorce modern stock from the ancestral form. As soon as the artificial conditions of cultivation and domestication were removed, their animals died or reverted to "type." Wild animals, on the other hand, had an inexorable tendency to depart from the original type; every step in their evolution was irreversible; the process could continue indefinitely.

Fully recognising the importance of this contrast between natural and domestic transformations, Wallace arranged his evolutionary arguments around it. At the head of his 1858 paper on organic change he stated that wild varieties had a "tendency to depart indefinitely from the original type" and proceeded to set out his reasons for believing this to be the case. The theory of evolution and the concept of a Malthusian selection of varieties were introduced as explanations—as the efficient cause—of this biological tendency. But the implications were nevertheless clear: Wallace was putting forward the idea of transmutation.

Here Wallace had an argument that, he felt, deserved consideration at the highest level. The paper, he decided, ought to be shown to Lyell. But who would introduce him to this famous geologist? A mutual acquaintance would have to act as intermediary. So in February 1858 Wallace anxiously sent his draft to the one man in England whom he knew to be interested in the species problem, with the request that it be forwarded to Lyell. In the middle of June it arrived at Darwin's home in Kent.

H.M.S. *BEAGLE* AND GEOLOGY

Darwin's path to the recognition of evolutionary changes in nature and his discovery of natural selection was in many ways similar to that taken by Wallace. Darwin drew on his travels on H.M.S. *Beagle* in the same way that Wallace drew on his expeditions through Amazonia and Indonesia. Darwin, like Wallace, was struck by the geographical variations of species and by the fauna of archipelagos; he too accepted the reality of evolution before finding a suitable agent for organic change. And both naturalists later found a source of inspiration in Malthusian hypotheses.

Yet Darwin's earliest work in natural history was conducted in a spirit entirely different from that of Wallace. Whereas Wallace had left England already converted to a developmental approach and determined to prove the truth of his views, Darwin struggled to show the superiority of Lyellian principles for an understanding of the structure and history of the earth. With the aid of a celebrated textbook, Darwin interpreted natural phenomena as if he possessed the eyes and mind of Lyell: "I cannot say how forcibly impressed I am with the infinite superiority of the Lyellian school of Geology over the continental. . . . I have always thought that the great merit of the *Principles* was that it altered the whole tone of one's mind, and therefore that, when seeing a thing never seen by Lyell, one yet saw it partially through his eyes."[29]

Darwin turned this enthusiasm to good stead. Travelling around the world in the *Beagle* during the years 1831–1836, he devised a general theory to account for the geological activity of South America and the Pacific. Most of Darwin's observations—including many strictly natural history observations—were recorded in order to supplement this one overriding theory. Indeed, it appears that Darwin's first philosophical interest in the origin and distribution of species came from his desire to emulate Lyell, from his wish to use biological arguments to support mainly geological proposals. Darwin's study of the biogeography of South America and the Galapagos, for instance, was carried out under this particular intellectual umbrella, and it is likely that his speculations about evolution grew out of the same set of ideas. For Darwin, then, the concept of evolutionary change was first generated as an integral part of another, more fundamental, geological scheme.

Darwin's interest in dynamic geology clearly had its origin in Lyell's *Principles*, volume I, a masterpiece of interconnected reasoning that attempted to demonstrate the constant activity of the earth's crust and showed how every phenomenon was part of a chain of events, in time as well as space. Lyell's earth was forever on the move. Activated by the movements of molten rock and pressure from internal heat, large sections of the earth's crust were pushed up or depressed relative to sea level. Elevation was part of a continuous fluctuation of the earth's surface and was always balanced at some place or another by reciprocal and compensatory subsidence. The whole system was not directed by any external force nor regulated from within, as might be the case with a gradually cooling earth, for example. Instead, it was self-balancing and self-contained. The earth was steady-state.

Lyell's vision of a steady-state earth lay at the heart of his "uniformitarianism." Balanced movements of crustal blocks seemed to explain the distribution of land and sea, the landforms and the topography of the present day, and were equally capable of explaining geological structures of the past. Although few of his contemporaries would have denied this role to elevatory and depressing forces, Lyell was unusual in emphasising such movements to the exclusion of other geological agents and was genuinely innovative in identifying causes for the phenomena. Lyell, who had thoroughly investigated the volcanic regions of the Auvergne and southern Italy[30] while formulating the ideas that were to become the *Principles of Geology*, used his knowledge of igneous rocks to account for elevation. Volcanic action was understood by him to be the release of pressure generated by molten rock constrained in vast subterranean reservoirs. If volcanic eruption was suppressed, liquid rock would be forced against the roof of its reservoir, so elevating the land overhead. And when the melted matter cooled—as all igneous material eventually did—the land above would remain permanently raised. His proposals therefore explained why volcanoes and mountain chains were often associated with elevated coastlines and why such regions were frequently shaken by earthquakes. He also explained the presence of granite at the core of modern mountains as a natural result of consolidation after elevation.

Much of Lyell's account was of necessity hypothetical. Darwin's achievement was to provide the first detailed information that supported and extended the arguments of the *Principles* and to supplement Lyellian dynamics with his own examples drawn from the southern hemisphere. Elevation could be used to explain many of the things Darwin saw. Thick alluvial deposits on the east coast of South America were best described as an elevated estuary, and Darwin suggested the fossil mammals found there must have been swept into the sea by Tertiary rivers, only to be covered with sediment and eventually raised above the surface. Rounding Cape Horn, he thought the Fuegian archipelago looked very like a row of submerged mountains and was gratified to find that the west coast of South America seemed to have been uplifted, step by step, from a state similar to that now exhibited by Fuegia. The evidence—drawn from a number of intellectual fields, but mostly based on successive beds of shallow-water mussels at heights varying from 60 to 230 feet above sea level—was more conclusive than even Lyell had hoped to find, and Darwin's work was accordingly greeted with particular interest in geological London. Further cross-country traverses and coastal surveys left no doubt in Darwin's mind that elevation had taken place exactly as Lyell had asserted.[31]

In 1835, as H.M.S. *Beagle* proceeded slowly up the coast of Chile, Darwin was able to draw practical conclusions about the causes behind such changes in the levels of land and sea. He and his fellow sailors saw a volcano erupt and soon afterwards experienced a major earthquake. Correlating this dramatic example of geological forces in action with his knowledge of the recent history of South America, Darwin came to believe that "earthquakes, volcanic eruptions, and sudden elevations on the coastline of the Pacific, ought to be considered as irregularities of action in some more widely extended phenomenon."[32] The more widely extended phenomenon he had in mind was the movement of the earth's crust, as suggested by Lyell. If the volcanoes of a mountain range were in some way connected deep below the surface, they would tend to erupt together or in sequence. Pressure not relieved by eruption might be dissipated through the fracturing and shifting of strata, causing earthquakes. At a less intense level of activity, molten rock would be constantly injected into the basal zones of mountain chains, thereby lifting the ground level "slowly and by little starts." The whole interrelated system effected a change in the level of one of Lyell's crustal blocks.

This train of reasoning, eventually set out in 1837 and 1838 in papers to the Geological Society and, later on, in the *Geology of the Voyage of the Beagle* (1842–1846), was in essence Lyellian. Although dealing with a single restricted area—South America—Darwin was making a powerful contribution to the elder man's scheme of things. Nor was Darwin oblivious to the global scenario presented by Lyell. It seems reasonable to suppose that his thoughts moved from South America to a larger theatre and from the problem of localised uplifting to a consideration of crustal movements on a worldwide scale. Since South America was prone to elevation, there must be areas equal in size that were likely to experience subsidence. Lyell's theory required such zones of elevation and depression to be adjacent, and common sense suggested that the place to find subsidence was in the sea. Darwin turned away from the continental landmass and surveyed the Pacific ocean. Before leaving the coast of South America he was convinced that this vast expanse of water concealed a subsiding sea floor. He was determined to search for the physical manifestations of just such a phenomenon.

Darwin found his evidence in the coral islands of the South Pacific and Indian oceans. With genuine scientific insight he realised that the shapes of coral reefs could act as indicators to the fact of subsidence. The coral polyp could not live at a depth greater than 120 feet, and it had always been said that reefs were confined to the shallow waters off mainland shores or oceanic islands. Lyell supposed polyps also perched on the rim of submerged volcanic craters, forming "lagoon islands." But if it was found that reefs went down a very long way and that all the coral below the 120-feet mark was dead—would that not be proof that the floor of the ocean had been gradually sinking and that the coral polyps had kept pace with this sinking by building the reefs up to the surface? If coral grows up and land goes down, there will be a succession of reef formations ranging from simple fringing and "barrier" constructions to more complex, landless "atolls." The particular type or class of coral formation depended on the extent and rate of subsidence that had taken place.

It is important to note that the actual theory of reef formation took second place in Darwin's mind to the question of subsidence. Conceived long before he saw a coral reef,[33] this elegant scheme served to support his contention that the course of geological history was marked by endless oscillations of the earth's crust. And the coral reef theory had another, more significant bearing on Darwin's general hypotheses: it showed him how markedly the living world was affected by geological processes and how delicately balanced were the relations between the organic and inorganic kingdoms. The living world was necessarily affected by recent changes in land and sea levels, and the case of coral polyps could be added to those of South American mussels and of fossil mammals as further evidence for a general theory of elevation and depression. Darwin, much more than Lyell, realised that biological arguments were crucial to the overall strength of a truly dynamic, historical, explanation of the earth.

THE GALAPAGOS ARCHIPELAGO

In September 1835, with these ideas fresh in his mind, Darwin sailed for the Galapagos. His interest in the archipelago was intense because it promised a new kind of scientific experience. The island group was of volcanic origin and, he thought, geologically recent; mountainous and studded with craters, the landscape was composed entirely of volcanic rocks in various stages of decomposition and at least two of the fifteen islands were covered with immense sheets of naked lava, signifying continuing geological activity. For Darwin, phenomena such as these meant one thing only: new land was being brought into existence. The geological forces that elevated the South American continent and fuelled the processes of mountain building were manifested here as a series of volcanic eruptions. At first hidden on the sea floor, these lava flows must have accumulated steadily, eventually emerging above the surface of the water as volcanic cones. Perhaps the process was accelerated by a general elevation of the sea bed, thought by Darwin often to accompany eruption. Where there had once been "unbroken ocean" there was now a constellation of new islands, ready to experience all that the world could offer. Here, in the Galapagos, Darwin could examine genuinely new land and judge the path of its subsequent history; he could study the effects of denudation and weathering and assess the rate at which base rock disintegrated into soil; he could virtually witness the transformation of the earth's raw materials into a fully diversified topography. This was the stuff of uniformitarian geology.

More important in the light of his later career, Darwin also got the chance to investigate the animal and plant life of the archipelago. Insular populations were fascinating objects at all times, and the Galapagos had a rich variety of endemic species. Here again it would be possible to see how animals and plants colonised new lands, how bare rock was clothed and peopled with living beings. In what way had these geologically recent islands acquired their species? Was it by immigration or by creation? Did new lands have new species? Darwin's interpretation of the geology of this region raised fundamental biological questions, the answers to which would surely reinforce his theory of crustal movement and expand its scope. He approached the Galapagean fauna and flora with a high degree of interest.

In conventional terms the Galapagean fauna belonged to the South American "centre of creation," but this could mean several things according to individual interpretation. Lyell, for example, understood species to have originated one by one, in different places within a general area. From these points of origin, species spread out and mingled with each other, making up a population that could be called a flora or a fauna. On the larger scale, floras and faunas could be considered as units in their own right and as bodies that expanded or contracted in size according to local conditions and topographical changes. Lyell explained the fauna of Sicily, for instance, as an outlier of the Italian population: the Italian fauna had spread from the mainland to Sicily when the latter was elevated from the Mediterranean sea floor. Sicilian species were not, therefore, created on Sicily but were natural imports from a neighbouring "centre of creation."

Following Lyell, Darwin could interpret the Galapagos animals as displaced members of some adjacent population. But it was more common to suppose that oceanic islands like the Galapagos, far away from the mainland and inhabited by endemic species, had experienced their own creative forces. They constituted tiny, satellite "centres of creation." The problem here, of course, was that it became impossible to explain the evident similarities between species of one centre and those of another without resorting to external agents like a Creator. Men such as Edward Forbes and Henri Milne-Edwards were compelled to invoke some kind of metaphysical resemblance or "representation" in order to satisfy this requirement.

Such idealistic reasoning did not appeal to Darwin. He was confronted by two mutually exclusive lines of argument: either the Galapagos species were the same as those of South and Central America, and consequently imported thence, or they were different from American organisms and ought to be thought of as independent creations. But the Galapagean species were neither one thing nor the other. They were the same and yet not the same. Faced with the existence of insular tortoises, mocking-thrushes, and finches built on the South American "plan," Darwin was forced to question the most fundamental of contemporary theories. He, it must be supposed, recognised that each hypothesis accounted for only part of the phenomena and attempted to make a compromise by fusing them together. He thought species had arrived from the continent more or less in the manner described by Lyell or by independent migration. The birds had clearly found different foods available on different islands, and it was reasonable to suppose the same applied to other forms of life. Isolated under different sets of conditions and through successive generations, species had adjusted themselves accordingly. The process gave the overall appearance of a special centre of creation.

It is doubtful that Darwin abandoned the doctrine of special creation in one dramatic step. There is some reason to believe the different forms of tortoises, for example, were interpreted by him as varieties of a single species, or, at the very most, as geographical races arising from the same secondary causes as varieties.[34] Certainly, he was surprised when the ornithologist John Gould informed him in January 1837 that Galapagean finches were not merely variants but true species.[35] It is quite possible that Darwin first considered American animals to have "sported" (varied) when they arrived on the Galapagos islets, the peculiarities then being preserved by continuing isolation. The mere fact of colonising new land—land that had been moulded by volcanic forces, lifted above the level of the sea, and left bare to the elements—may have been sufficient to disturb the normal pattern of reproduction in these living beings. There was no need for him to suppose that the present organisms were distinct *species*.

ORGANIC CHANGE

Nevertheless, at some stage during the voyage back to England, or very soon after his arrival there in October 1836, Darwin did take this momentous step. From the month of March 1836 he had constantly kept a red notebook by his side and filled its pages with field observations; from October these were replaced by speculations about geology, species, and geographical distribution. He had decided that species change and give rise to other species.[36]

In retrospect the biogeography of South America seemed significant, and Darwin noted how the spatial distribution of the rhea was comparable to the chronological relationships of past and present mammalians. The common ostrich (as he called it) was replaced in the southern parts of the continent by another, closely allied species, the *Rhea darwinii*; the two were taxonomically related yet mutually exclusive in area. The same could be said about the Tertiary fossils discovered in the Pampas and the present occupants of the region: the two sets of mammals were plainly related and yet mutually exclusive in a chronological sense. The processes that encouraged spatial differentiation must also have stimulated changes over time: whatever it was that made one modern species devolve from another would also account for the action of evolution in the past. Darwin had come to see the full meaning of Lyell's dictum, "As in space so in time."

In his red notebook Darwin seemed confident that isolation was the basic factor behind specific change, and his search for further information took him on a circuitous path from islands and geographical barriers to animal behaviour and breeding habits, to mechanisms of generation and back again to islands. His geological theories, now being described in detail to the Geological Society of London,[37] became inextricably mixed up with these new ideas about organic transformations and were henceforth inseparable. Elevation and depression of land relative to sea level were considered by him to be excellent devices for the introduction and removal of topographic barriers, for the creation or destruction of oceanic and coastal islands, for the preservation or extinction of species. Darwin, it must be said, had no fixed ideas about how such concepts should be articulated together, but he made it plain that the notion of organic change was firmly associated in his mind with the idea of islands, real or metaphorical.

During this richly thoughtful period Darwin opened a series of notebooks, now famous for their extended struggle with the species problem. From July 1837 to September 1838 he puzzled over the agents that might stimulate variation, divergence, and adaptation, the causes of geographical distribution patterns, and the factors that led to extinction.[38] Darwin was interested in everything and had so many ideas that the work—the concept of descent with modification—took on ever widening horizons and involved more and more fields of inquiry. He searched everywhere for a mechanism for evolution. That search, as is well known, was triumphantly concluded in September 1838, with Malthus's axiom that populations will outstrip food supplies unless checked by some means or another.[39] At last, he had "a theory by which to work."

But Darwin, more than anyone, recognised that the discovery of "natural selection" was only the start of a long programme of research and reflection. The substantiation of this flash of genius would require a well-developed theory and a comprehensive account of evolutionary change in nature. All the topics so briefly touched upon in his notebooks would now have to be thoroughly explored. Unlike Wallace, who found that the idea of natural selection completed what was already a mature and expansive hypothesis, Darwin felt his life's work was just beginning.

The years between 1838 and 1842 were busy ones for Darwin. While completing the publication of his geological papers, supervising the composition of the *Zoology* of the *Beagle*, watching over the identification of specimens, and writing his own *Geology of the Voyage of the Beagle*, Darwin took on official duties in the Geological Society, married, and set up house, first in London and then at Down in Kent. His time was also filled by the collection and collation of information intended to flesh out the bones of his discovery. He developed, in three years, an integrated scheme of geological and biological change, first described in a preliminary *Sketch* of evolutionary theory in 1842[40] and in greater detail in an *Essay* of 1844.[41] All his old interests—islands and the subsidence and elevation of landmasses—were retained and placed at the hub of these maturer arguments. Geological transformations were to be an integral part of the theory of "descent with modification."

In his 1842 *Sketch* Darwin focussed on islands as the primary agent in stimulating organic change and as the mechanism by which newly modified forms could be kept isolated without the deleterious effects of belonging to a large and interbreeding population. He envisaged an island or archipelago already peopled with aboriginal species or occasional accidental migrants. Drawing an analogy with the effects of domestication on the reproductive systems of farm animals and cultivated plants, Darwin suggested that wild species were similarly affected, that they were "unsettled" when dislocated and subjected to alterations in their conditions of existence. Geological change, for example, would upset the "constitution of some of the insular species" and precipitate the appearance of structural variations in the offspring. In particular, the act of crustal elevation would reveal tracts of land hitherto unoccupied by terrestrial living beings and so provide local species with dramatically new environments.[42] Here, they would vary unchecked. Natural selection would go to work, preserving favourable variations and weeding out the unfit. Prolonged isolation on the one islet, coupled with the fact that such regions generally had only a few species to fill all the available spaces, would lead to the production of many new species. For these reasons, contended Darwin, "an island would be a far more fertile source, as far as we can judge, of new specific forms than a continent." The Galapagos archipelago, frequently cited in the 1842 *Sketch*, was clearly uppermost in his mind. In the Galapagos, he assumed that an original importation of organisms such as the South American finch had varied at random and had been forced by natural selection to adapt habits and structure to the "places" left free in nature.

But there was more than this to Darwin's scheme. Applying his knowledge of the processes by which large continents came into existence, he reasoned that continual volcanic activity would lead to more and more islands being produced in an area such as the Galapagos. Eventually, continued elevation and eruption would produce a line of mountains surrounded by intermediary low-lying regions where the individual islands had fused together. The animal and plant life native to each islet therefore had the chance to expand its range and mix with new associates; the flora and fauna would intermingle until such time as partial subsidence separated the irregular landmass into a set of discrete islands once again. The consequences of these alterations in level were plain: species would be jumbled into new associations and brought into new relationships with their peers; the conditions of their existence would once again be upset and they would vary and change; selection and evolution would proceed apace.[43] Rising land therefore encouraged speciation, and fully elevated land, dispersal. Subsiding land, on the other hand, tended to render species extinct (at least locally), because the process removed many different kinds of habitat at the very time that it was hardest for animals and plants to migrate. Alpine plants, for example, would find nowhere else to go if the tops of mountains subsided into warmer zones and would speedily die. Darwin's scheme was simple: with successive changes of level species evolved, dispersed, and became extinct.

If this process of oscillation was repeated enough times—and Darwin made the point here and elsewhere that elevation and subsidence did go on and on, in just such a fashion—then all the multifarious relations between the beings of separate countries could be explained. Speaking generally, the crust of the earth went up and down, altering insular topographies to continental ones and vice versa, and species changed or migrated according to their location on an island or a united landmass. Darwin had a picture of a dynamic flux of living beings—beings that evolved, dispersed, and died—which superimposed itself on the equally dynamic movements of the earth.

Darwin's vision was necessarily somewhat different from Wallace's. Although Wallace was deeply influenced by Lyell, by the concepts of uniformitarian geology and his emphasis on a dynamic earth, Wallace did not create a theory that depended on Lyellian changes as Darwin's did. Darwin needed movements of the earth's crust to provide alterations in physical topography; only these were sufficient, he thought, to "unsettle" the constitution of existing species and thereby encourage them to vary. The genetic mould, as it were, had to be broken by definite changes in the physical or social environment before the organism would sport and before natural selection could go to work. With this emphasis on external "unsettling" agents, Darwin was fulfilling the obligation placed on him by the use of artificial selection as an explanatory model. The parallel between the "artificial" world of domestic animals and the situation in the wild was, of course, central to Darwin's argument,[44] yet it has been little noted that this parallel extended beyond the action of selection to embrace the actual causes of variation as well. Darwin had drawn certain conclusions about the reproduction of domestic and cultivated organisms and was forced to seek out forces and agents in nature that would mimic the effects of domestication. In a nutshell, he had decided that animals and plants removed from nature suffered radical alterations in the conditions of their existence, a process that inevitably stimulated extreme and continuous variation in the offspring. The breeder then selected the "right" variety for future propagation. This interpretation of events required a parallel force in nature that altered the conditions of existence of wild species and so precipitated variation. Darwin found one in the system of geological movements that had dominated his thoughts for nearly ten years. His geological theories were as fundamental to *this* theory of evolution as the crucial idea of selection itself.

Wallace, on the other hand, believed there was no parallel between domestic and natural transformations. The two processes were, for him, radically different. Domestic varieties were attuned to an artificial world, and therefore liable to revert to less specialised forms when these false conditions were removed. Natural varieties, conversely, were selected because they were best adapted to the *real* world—they could not regress without dying. Wallace's domestic breeds had to revert or die; his natural stocks had to press on to avoid the same fate.

Wallace rejected an analogy between the two kinds of organic transformation because he did not consider the problem of the origination of variation to be important. Familiar with over eight thousand species of insects alone, Wallace had found nothing in nature to contradict the principle laid down by Robert Chambers that reproduction was always accompanied by some minor variation. For Wallace, variation was an integral part of the reproductive process and, as such, required no special line of explanation or argument. It was taken as given. By contrast, however, Darwin believed in 1842 that variation was in some way unnatural—or, rather, was induced by forces extraneous to the reproductive system. Variation was held to be stimulated by particular external actions or activity, which changed the pattern of propagation from simple replication of species to one where unbridled variation took place. On this critical point Wallace and Darwin were dramatically opposed.

So, despite the remarkable parallels in the prehistory of Darwin's and Wallace's theories of evolution it must be acknowledged that they were, in the first instance, significantly different—different in their use of the idea of artificial selection; different in the way they thought variation arose in populations; and different in considering the level at which evolution was supposed to be taking place, Wallace looking at varieties where Darwin thought only of individuals.

Such a distinction would not perhaps be important if it were not for the fact that Darwin, over a period of about twelve years, gradually surrendered the greater part of his 1842/1844 theory of transmutation in favour of a scheme that was altogether more Wallacean in tone. Darwin relinquished his insistence on external changes in living conditions as the cause of variant offspring. He moved away from the consideration of individual variants to a position in which varieties (the taxonomic category) were thought highly significant to the evolutionary process. And in so doing he sacrificed his system of geological tectonics, reduced the role of geographical isolation, and introduced a compensatory principle—the principle of divergence. By 1856 or so, he was in command of what seems like a different theory.

Why did Darwin do this? It certainly had little to do with Wallace, since the changes were all but completed by the time Darwin became aware of Wallace's 1858 Ternate exposition. More positively, it did have something to do with Darwin's continuing work in natural history and, specifically, the extension of his studies of geographical distribution. Darwin became, as it were, a geographical botanist.

EIGHT

Darwin's Distribution

By the year 1844 Charles Darwin had already devised and presented a comprehensive account of evolutionary transformation. Nevertheless, as is well known, he then went on to make quite substantial alterations to his theory. Later versions not only covered a much wider range of topics but were put together with far greater philosophical insight, reflecting many years of literary and practical research. Yet there were other, more fundamental alterations than these. The later versions also represented a complete turnabout in Darwin's views on the mode of evolution. Dov Ospovat's careful study of Darwin's work through the 1840s and 1850s shows how he abandoned the geological substructure of his 1842 *Sketch* and 1844 *Essay*, changed his mind about the origins of variation, and created an additional principle—divergence—to act in concert with natural selection.[1] Such dramatic reconstructions had only just been completed when Darwin received Wallace's letter from Ternate in 1858. By force of circumstance this new version, not the earlier theory, was the one presented to the public in November 1859. The *Origin of Species* was more rushed than has ever been suspected.

The conceptual differences between the *Essay* and the *Origin* are attributable to one crucial discovery which compelled Darwin to amend several vital lines of argument and to reevaluate the overall structure of his original theory. Through an extended study of living and fossil barnacles, carried out between 1846 and 1854, he came to believe that the natural world contained a great deal of spontaneous variation. Contrary to his written opinion of 1844, variation seemed to occur despite the absence of unsettling agents such as geological change. Wallace already knew this, but Darwin had to discover it for himself the hard way. Like Wallace, Darwin immersed himself in the complexities of taxonomy, which led him from the examination of a single species of burrowing cirripede to an eight-year investigation of the whole group. This thorough survey taught him—as no amount of reading or reasoning could have done—that variations constantly arise in the offspring of ordinary species without the action of external forces. Variation was a natural consequence of reproduction.

Although his fundamental volte-face did not originate in Darwin's continuing work in geographical distribution, it is here that the changes forced on him are most clearly seen. Where before he had confidently employed a system of geographical isolation as a major element in evolution, he now felt obliged to replace it with some alternative mechanism; where his studies of migration and dispersal once included a belief in changes in sea level and continental extensions, he now emphatically rejected land bridges and the like; where before he had been only marginally interested in the relative numbers of species and genera, after 1854 he adopted the whole question with enthusiasm and carried out a complicated and wide-ranging investigation into botanical arithmetic. These arithmetical exercises supplied the context for his "principle of divergence"—or, more precisely, raised problems that could only be solved by the intervention of an extra "force" in evolutionary theory. In no small way, the development of Darwin's ideas of "descent with modification" in the interval between 1844 and 1858 hinged on his work in geographical distribution.

MIGRATION AND LAND BRIDGES

One of the most important intellectual issues for Darwin between 1842 and 1846 was the question of geographical range. His first (1842) theory of evolution stipulated that species were capable of extensive migration, that organisms could freely move to occupy areas made available through topographical change. Species that could not reach insular land had little chance of evolution under Darwin's scheme; species that could not disperse from their island origins had few opportunities to influence life's history. The *Sketch* and *Essay* therefore assumed that animal and plant dispersal was a common occurrence, dependent mainly on the elevation of intermediary land but also on their own inbuilt powers of migration. The crux of Darwin's system was the proposition that species could spread virtually all over the world, given plenty of time and no physical or ecological barriers in the way. They spread "as far as barriers, the means of transportal, and the preoccupation of the land by other species, would permit."[2]

This notion of wide-ranging organisms kept Darwin fully occupied for the greater part of 1843 and 1844. His early letters to his new acquaintance Joseph Hooker at Kew were full of questions about the comparative range of species. On 23 February 1844, while researching his *Essay*, he wrote to ask if there was any botanical counterpart to the suggestion that animals might inherit the ability to spread far and wide. Was there any relation between the range of a species and its genus? "In discussion with Mr Gould," he explained, "I found that in most of the genera of birds which range over the whole or greater part of the world, the individual species have wider ranges, thus the Owl is mundane, and many of the species have very wide ranges. . . . Is it not so with Cryptogamic plants?"[3] In other words, did the species of a well-dispersed genus have wide ranges themselves? Discounting the evident fact that any genus might have a far-flung empire through the mere juxtaposition of very many small ranges, Darwin wanted to know if all the species of a group such as voles or mice, for example, had wide ranges each in its own country. He was searching for confirmation that one species with the power to spread out over the globe would give rise to descendent species, every one capable of doing the same.

This and similar questions were fired off in rapid succession in letters to Hooker and other naturalist friends. Darwin asked Hooker if he thought insular species were more variable than their continental counterparts, another important aspect of his current theory of evolution.[4] Bory de Saint Vincent had said so in his *Reise nach den Kanarien-Inseln* (1805). If true, it would be a bonus for Darwin. Unfortunately, Hooker thought insular species, if anything, were the very model of stability, showing little tendency to vary.[5] What, then, did Hooker think of aberrant species and genera; did these forms have a more restricted range than "typical" species, as William Swainson suggested? Had he any conclusions yet on the Galapagos flora? What did Hooker think of Edward Forbes's account of arctic and alpine distribution patterns?—Darwin confessed himself nonplussed at being forestalled in such a fashion. To read through this early correspondence is to catch a glimpse of Darwin's mind at its best, vibrating with scientific endeavour and concern.

His interest in geographical range revealed itself in other ways as well. One topic which arose after the *Essay* of 1844 and occupied him more and more through the end of the 1840s and into the 1850s was the power of individual organisms, adults and offspring alike, to migrate. On the theory worked out in his *Essay*, every flora and fauna with even a slight affinity must once have been part of the same geologicobiological system. Assuming that an aboriginal organism possessed the capacity to range widely over the lands that now form, say, Europe and North America, the present-day faunas of both ought to include species descended from the common ancestor but no longer exactly alike. To avoid the unwelcome suggestion of an enormous continent in the middle of the Atlantic, Darwin attributed the link to the sheer dispersive powers of the animals themselves. In the southern hemisphere, for example, the marsupials of South America were clearly related to those of Australia, yet there was no evidence for a landmass big enough to bridge the gap between the two. It was easier to fall back on the ancestral opossum's ability to shift for itself. But how?

To show that animals and plants could travel much further from their "centres of distribution" than was usually believed, Darwin undertook a delightful series of practical experiments. He kept the seeds of many different plants, some specially chosen because they lived on islands, in salt-water tanks in his cellars at Down in order to prove that they could survive for a certain number of days whilst being whirled about the oceans of the globe. "To my surprise," he wrote, "I found that out of 87 kinds, 64 germinated after an immersion of 28 days, and a few survived an immersion of 137 days."[6] All these seeds, however, had sunk to the bottom of his phials of seawater, which, he told Hooker ruefully, did not say much for their chances at sea. Dehydrated seeds, on the other hand, did float for a limited number of days, prompting Darwin to re-create the circumstances in which "floods might wash down plants or branches, and these might be dried on the banks, and then by a fresh rise in the stream be washed into the sea." He dried the stems of fruit-laden plants and recorded their survival in seawater. Hazelnuts, for instance, floated for 90 days, asparagus for 32, both successfully germinating at the end of their ordeal. Darwin used this information to compute, with the aid of Johnston's *Physical Atlas*, the distance over which his seeds might have travelled. Assuming an average rate of 33 miles per day for the several Atlantic currents, he calculated that seeds floating for about 28 days would be carried across 924 miles of sea.[7]

Other more irregular means of transport were also included in the investigation. Seeds and larvae might travel with birds, alive or dead, attached either to their feet or their feathers, or else inside the gizzard and digestive tract. Darwin's children were fascinated by his suspension of a pair of duck's feet in an aquarium stocked with freshwater snails, to "represent those of a bird sleeping in a natural pond." Some recently hatched snails obligingly crawled onto the feet and clung there firmly while Darwin tried to shake them off. These resilient molluscs survived in damp air for a period of 12 to 20 hours. "In this length of time a duck or heron might fly at least six or seven hundred miles, and would be sure to alight on a pool or rivulet, if blown across sea to an oceanic island or to any other distant point," wrote Darwin.[8]

Trivial as his experiments might seem, they proved highly significant for the theory of evolution, because the results established beyond doubt that species were capable of spreading far more widely than had hitherto been supposed. Darwin was the first to provide actual quantitative details, thereby adding considerable weight to his assertion that animals and plants were constantly on the move, ever ready to press into the slightest gap in the ecological carpet. His results also confirmed the importance of individual migration, independent of topographical or other geographical changes. He did not by any means abandon the idea of topographical alterations, which at that time remained firmly at the centre of his speciation theory. But his practical experiments did reduce the importance of geology and geography as a prime factor in every stage of evolution.

Furthermore, as Darwin was quick to point out, he was now better placed than most naturalists to refute the doctrine of multiple creation. Those who believed in single creation (as did Forbes and Hooker) were obliged to explain every set of disjunct species by supposing their separate habitats had once been connected together or open to communication with each other. Thus Forbes explained the existence of the strawberry tree in Ireland and Portugal by postulating a great mid-Atlantic land bridge, over which this tree and a species of nudibranch slug had migrated during the Miocene period.[9] If Forbes's geological evidence was disputed or, as seemed highly probable, was found to be false, then where would his argument for single creation be? He would have to admit the possibility of multiple creations. Darwin had a second line of defence against such geological criticism with his emphasis on accidental and occasional means of transport. Even if the topographical configuration of the globe was not conducive to ordinary dispersal, Darwin's accidental means could still account for the appearance of a particular organism in two disparate spots and so preserve the thesis of single origins for species. "If those botanists who have especially and lately attended to subject," he noted, "such as Alphonse de Candolle, Joseph Hooker and Edward Forbes, that plants cannot be believed to have crossed except on solid land from continent to Madeira, from Antarctica to New Zealand, and if we are forced to give up continents, multiple creation must be admitted."[10]

Philosophers to a man, Darwin's friends appeared to have restricted themselves to a dangerously narrow train of thought. But Darwin had already learned to his cost not to place too much faith in the "principle of exclusion" used by naturalists to dismiss as impossible any theory not agreeing with their own. In 1839 he had rejected outright an explanation of the "parallel roads" at Glen Roy—shelflike structures around a Scottish glen—which did not rely on the theory of marine erosion. To his dismay and embarrassment, Agassiz subsequently came up not only with a viable alternative, but one that was far more plausible.[11]

So Darwin pressed ahead with the idea of organisms spreading far and wide of their own accord, over and above the usual extension of range that might accompany local geological changes. The more he worked on occasional modes of transport, the less he needed structural alterations in the surface of the earth to account for distribution patterns. Hence in 1846, and consistently thereafter, he began to reject the idea of vast continental extensions and land bridges in favour of the individual organism's capacity to migrate. Such great changes in the earth's structure were unnecessary—and in the present state of geological science, mostly unverifiable—when there was an alternative mechanism in occasional transport: "I must maintain that I have never let down or upheaved our mother-earth's surface, for the sake of explaining any one phenomenon, and I trust I have very seldom done so without some distinct evidence. So I must still think it a bold step (perhaps a very true one) to sink into the depths of the ocean, within the period of existing species, so large a tract of surface," he told Hooker in 1846.[12]

Nearly ten years later Darwin made his position crystal clear. In a splendid display of indignation, he complained to Lyell that the latter's disciples were elevating and depressing huge sections of the earth's crust with unscholarly abandon. "Here poor Forbes made a continent to North America and another (or the same) to the Gulf weed; Hooker makes one from New Zealand to South America and round the world to Kerguelen Land. Here is Wollaston speaking of Madeira and Porto

Santo 'as the sure and certain witnesses of a former continent.' . . . And all this within the existence of recent species! If you do not stop this, if there be a lower region for the punishment of geologists, I believe, my great master, you will go there."[13]

Where was the evidence for such recent land bridges and extensions, demanded Darwin. The hypotheses of Forbes, Hooker, and Wollaston were based on only a few species with split ranges or on general biological affinities between the regions in question; there seemed to be little *geological* evidence for former continents amid the Atlantic or Pacific oceans. Lyell's theory of elevation and depression had been misinterpreted. Oscillations in the level of continental landmasses, he pointed out, were not the same as great projections across the oceans of the world: "I do not for one second doubt that very large portions of all these continents have undergone *great* changes of level within this [current geological] period, but yet I conclude that fundamentally they stood as barriers in the sea, where they now stand; and therefore I should require the weightiest evidence to make me believe in such immense changes within the period of living organisms in our oceans."[14]

Moreover, the supposed continental extensions did not provide an adequate explanation for all the phenomena of distribution. If oceanic islands had once been connected to the nearest mainland, as argued by Forbes and Wollaston, why did the islands not have mammals or frogs, which should have been able to cross all the way on dry land? Why was there an absence of certain great groups of insects on Madeira, or of acacia trees in New Zealand? To be sure, remarked Darwin, the hypothesis of accidental dispersal could not account for all these exceptions, but it was still more plausible than the theory of continuous land. His objections boiled down to the conviction that "we are extremely ignorant of means of dispersal." Dispersal was the right road to take, in his opinion; continental extensions were merely a blind alley.

By the middle of the 1850s, at the height of the debate with Lyell, Darwin found yet another reason to reject the idea of continental extensions. But this was only one aspect of the complete change-round in his ideas that followed his investigation into the order *Cirripedia*. The study of barnacles, begun in 1846 and only given up in 1854, when he had examined the whole order in minute detail, had a dramatic effect on the course of Darwin's thoughts about the actual mechanism of organic change. It caused him to pass over the geologicobiological model of speciation, described so fully in the *Sketch* and *Essay* and the source of his researches into dispersal and geographical range, and led him on to search for another explanation that would incorporate important new discoveries. In the process, Darwin's appreciation of geographical distribution changed focus and intensified in depth.

A NEW APPROACH TO VARIATION

For reasons best known to himself Darwin devoted eight years of his supposedly precarious life to an extensive and all-embracing account of living and fossil cirripedes. Contrary to what he had expected in 1844, the cirripedes revealed a great deal of variation in the natural world. In the *Sketch* and *Essay* he had insisted that very little variation was seen in a "wild state" and had repeated over and over again that "most organic beings in a state of nature vary exceedingly little."[15] He had relied instead on geological and geographical change, either directly or indirectly (the latter by precipitating organisms into different social patterns), to "unsettle the constitution" of wild animals and plants. The "unsettling" agents were presumed analogous to the effects of domestication on the reproductive systems of organisms under any sort of cultivation. By 1854, however, Darwin was convinced that organisms in their natural state really did vary without any such "unsettling" forces. In a letter to Hooker, he claimed: "You ask what effect studying species has had on my variation theories. . . . I have been struck with the variability of every part in some slight degree of every species."[16]

Such a discovery of course weakened the arguments of his *Essay*, in which he had drawn a close analogy between selection in the wild and under domestication. A change of circumstances in either case was assumed to lead to a "certain plasticity of form," with the reproductive process producing variant offspring subject to the forces of natural selection. The crucial point, according to the *Essay*, was that variants arose only when the generative process was disturbed. Now, however, Darwin knew that varieties popped up in the wild without rhyme or reason, thus invalidating the central analogy of his thesis. In September 1854, as soon as he had completed work on the barnacles, Darwin turned to take a fresh look at his thinking in the *Essay*.

As might be expected, he focussed on the problem of variation. How did variation occur if there were no geographical or geological changes? Was tectonic activity entirely superfluous to the theory of evolution? Something else was obviously required to unsettle the constitution of species, but what? And without geographical isolation, formerly an essential part of the theory, how were "good" variants maintained in a population? Darwin had much to think about.

During the following years he struggled to find answers to at least some of these important questions. The origination of variant offspring was perhaps the most problematic and, in a way, was never fully resolved before the beginnings of twentieth-century genetics. Every animal and plant, he proposed, engaged in sexual reproduction at some stage in its life cycle, the act of cross-fertilisation maintaining the vigour and fertility of subsequent offspring. Coupled with slight alterations in the parents' conditions of existence, Darwin suggested there would be corresponding changes in the structure and habits of the next generation. Variations therefore appeared as a natural consequence of sexual relations.[17] Darwin, of course, did not differentiate, as we do, between the process of reproduction and the manufacture of variant offspring; between the act of fertilisation and the procedures of genetic recombination. He assumed that anything impinging on reproduction must necessarily affect generation and the resultant young. Where he had once emphasised the action of agencies external to the reproductive system, he now believed in internal factors and the fundamental role of sex.[18] This innovation was one of the great intellectual achievements of the interregnum between the *Essay* and the *Origin*.

Darwin also managed to find a solution to the second crucial problem that had emerged from his rereading of the *Essay*. He needed something that would preserve organic novelties once they had arisen, something that would permit the transmission of biological changes through many generations. Ordinary reproduction among the small populations typical of island faunas and floras previously had served this function, but now Darwin was forced to consider larger groups of individuals among whom continuous interbreeding seemed likely to swamp advantageous variants. In response, he constructed the theory of "pangenesis," a brave attempt to frame an explanation for the transfer of hereditary resemblances, the inheritance of some acquired characteristics, and the phenomena of atavism and regeneration.[19] Unable to distinguish between the hereditary material and the visible properties (genotype and phenotype in today's terminology) of the organism, Darwin suggested that every organ and structure of the body gave off minute particles called gemmules which ultimately found their way to the reproductive system, where they were assembled and passed on to the offspring. Modifications in the adult organism would then be transmitted to future generations. Because the system was based on a quasi-particulate inheritance, Darwin hoped to avoid the swamping effect attendant on larger populations. In this he was not altogether successful.[20]

But the issue that perplexed him most was the mode of evolution itself. On 9 September 1854 his journal records that he "began sorting notes for species theory,"[21] and turned to the biggest question that his revised version of the *Essay* would have to answer—the problem of how a superabundance of variation in the wild would affect his previous statements about the origination of species, and how speciation and extinction occurred when constant geological or geographical changes were no longer a necessary part of his theory. Why, if there was a great deal of variation in nature, did species become extinct? Surely their variability ought to permit a wealth of modifications to suit changing environments? And what of speciation? If variation was not tied to insular species, in what sort of species did variants appear? Perhaps by studying varieties and, if possible, finding regularities in their appearance in certain genera or species, Darwin could arrive at a fuller understanding of evolutionary change. If there were any rules to variation he wanted to know what they were.

At the beginning of 1855 Darwin began a series of simple arithmetical studies, aimed at identifying numerical correlations between species and varieties and the genera to which they belonged. He attempted to demonstrate that large genera (that is, genera with a large number of species) presented an above-average number of varieties, or, alternatively, a great number of forms in between varieties and species, such as "incipient species" and "closely allied species." Not content to rest his case on a few and perhaps eccentric examples, Darwin set out to explore the whole issue, twisting and turning to look at it from as many angles as possible, working his way through a pile of printed catalogues and an enviable abundance of fresh ideas. He made a careful calculation of just about every relationship he could think of that included large genera and varieties somewhere in the proposition and was gratified to discover that the computations all pointed in the same direction: genera with many species did indeed possess a greater proportion of taxonomic varieties, as well as "incipient" and closely allied species. In every one of the twelve or so volumes that Darwin examined this relationship held true. He worked his way through the trusted Floras in his collection, such as Henslow's *Catalogue of British Plants* (second edition, 1835), Hooker's *Flora Antarctica* (1844–1847) and *Flora Novae Zelandiae* (1853–1855), and Hooker and Thomson's *Flora Indica* (1855), in addition to works given to him by his recent acquaintances, the American botanist Asa Gray and the irascible British naturalist Hewett Cottrell Watson.[22]

Under the influence of Alphonse de Candolle's substantial *Géographie botanique raisonnée*, which he read soon after its publication in August 1855, Darwin also expanded his own survey of the incidence of varieties to include Candolle's ideas about geographical range and the frequency of individuals. However, where Candolle used families (or what he called "natural orders") as the base of his calculations, Darwin preferred to use genera.[23] Genera with many species, he found, were often not only mundane but possessed many individuals in each constituent species. Such a correlation might perhaps have been arrived at by a priori reasoning, but because of the obvious pitfalls Darwin wanted to establish the point to his own satisfaction; he was well aware of several instances to the contrary. There were cases of genera with only one or two species spread very widely over the earth, such as the tulip tree, and others of small genera with very prolific species, such as the earwig or penguin. Yet working through his Floras, following the example of Candolle, Darwin confirmed a slight but consistent tendency for the two characters of a great geographical range and a multiplicity of individuals to appear in the larger genera. As to the latter, for instance, Darwin examined the ubiquitous species in Alexandre Boreau's *Flore du centre de la France* (1840): the genera with common species (as denoted by the symbol C.C.) possessed an average of around 4 species per genus, whereas the remaining genera without such common species presented only 1.8. Clearly, the abundant species tended to occur in the larger genera.[24]

This extended survey of large genera and varieties produced several notable conclusions in line with his current theories. Darwin's botanical arithmetic, for example, explained much that was problematic in geographical distribution. He made immediate use of these results in his chapter on distribution for the "big species book," *Natural Selection*,[25] composed during the earlier part of 1856 and revised occasionally thereafter to include new information, such as that derived from Asa Gray's "Statistics of the Flora of the Northern United States," published in September 1856. In the belief that it was the big groups in nature that were most widely spread, Darwin could explain the origin of closely related yet geographically mutually exclusive "representative" species by asserting that as a species spreads out over a great area it will meet with different conditions, which stimulate local adaptations. The result would be a series of taxonomically related neighbours. The more representative species in a genus, the greater the size and geographical range of the group. By determining whether they contained either "discrete" or "close" species, dissimilar or similar relations, Darwin could also differentiate between genera that were small because of extinction among the ranks and those that were small because they were at the beginning of their life. The former indicated an "old" genus that was gradually dying out through the extinction of first one and then another of its species, rendering the remaining forms rather distinct from one another; the latter implied a "new" genus that was varying and producing many species all on the same plan. "Old" genera—with taxonomically distinct species—usually had a sporadic or disjunct distribution pattern, because the existing forms were remnants left behind when a large and correspondingly widely spread genus had died out. By contrast, "new" genera were composed of contiguous specific ranges. Furthermore, Darwin was able to explain why some organisms were rare and others abundant, depending on their occurrence in small or large genera, although here he was the first to concede the existence of many additional factors that allowed, say, a plethora of individuals in small (and supposedly "old") genera like those of the earwig or platypus.

Clearly, these statistical findings had a direct and valuable impact on Darwin's account of geographical distribution.[26] The study of botanical arithmetic was influential in other respects as well, leading him ever deeper into the problems of speciation and extinction and revealing tantalising clues to the processes of variation and evolution. Most important of all, his arithmetic stimulated him to construct a new explanation for specific change, almost entirely based on these ideas about large genera.

ADAPTATION, ISOLATION, AND EVOLUTION

It seemed to Darwin that larger genera ought to be the ones "best adapted" to their countries. Variation, it will be remembered, was the raw stuff of speciation as far as he was concerned. It was on variants that natural selection worked to produce forms well adapted to their surroundings, and an accumulation of such adaptations "made" a new species. If, therefore, most varieties occurred in the larger genera, these taxa should be the "best adapted" to their local environment, and small genera should be the least adapted.

Darwin tried to explain this by relating the quality of adaptation to some physical attribute that genera might possess, such as the capacity to range widely over diverse terrains. In a private note to himself, written in 1855, he observed that widely spread species seemed to exist in geological time for a longer period than most species, a fact he attributed to their "high" adaptative powers. "It is," he wrote, "that wide spread [range] shows that [the species is] best adapted and therefore survives longest."[27] And again, thinking on the reverse case of small genera that were strictly local in geographical extension: "Genera with few species show that those peculiarities which the species have in common are not so well adapted to the country [being] inhabited, as those genera with many species; and hence they do not range so far."[28] Darwin associated the size of the genus with its potential and actual adaptation, measured in this particular instance by geographical range but just as easily gauged by any of the parameters he had introduced into his botanical arithmetic—the number of varieties, for example, or the abundance of individuals. Adaptation was the factor that linked them all together.

Darwin used this notion of adaptation to explain how his numerically large genera eventually turned into smaller ones, and vice versa. He called this the "coming in and out" of genera in the history of life, and it signified the process of the "birth" of a "new" genus out of an existing group of species and the latter's subsequent decline into an "old" genus. For Darwin in 1855 and 1856 this process was not so much a splitting up of one great spectrum of species—as in his principle of divergence—as a breaking off, where one or perhaps two well-adapted species devolved from the parent genus. The parent genus was thought to die out as the break-away forms flourished and grew. Since such notions are markedly different from those later put forward in the *Origin* under the label of "Divergence of Character," it is important to understand precisely what it was that Darwin meant at this particular time, during 1855 and 1856.

He appears to have envisaged that a large, prolific genus—one which also spread widely—would have some species at the edge of its range exposed to "many conditions and several aggregations of species." They were exposed to the elements that encouraged struggle and competition. Should such a species vary, as species belonging to large genera were likely to do, he continued in another private note, "it may be selected to fill some new office, and mere chance would determine the origin in a large genus of some new and good modification."[29] To paraphrase, the introduction of a new and favourable adaptation to a satellite species at the edge of a large geographical range would lead to the formation of a new line of development and, ultimately, to the rise of a new genus and the demise of the old.

Certainly this was how Darwin explained his ideas to Charles Lyell, when pressed to give details of the theory of transmutation in 1856. Lyell recorded the conversation in his "Scientific Journals": "Genera differ in the variability of the species, but all extensive genera have species in them which have a tendency to vary. When the conditions alter, those individuals which vary so as to adapt them to the new circumstances, flourish and survive while the others are cut off."[30]

This view, of course, was not completely unfamiliar to Darwin. He had used such an explanation as early as the notebooks and consistently ever since. Ernst Mayr and others have rightly emphasised the significance of geographical isolation in this respect and Darwin's use of it as the process by which divergence of character could take place.[31] But here it can be seen how deftly Darwin wove new findings, such as the statistical evidence for variation, into the fabric of his earlier thought, and how he expanded and embroidered his ideas as fresh information became available through his own researches. Previously he had relied on islands and other geographical barriers, and on the relations between land and sea, to provide an isolating mechanism that would allow species to diverge from the original genus. Now, however, he believed large genera ranged widely and were prone to vary. Some species were likely to meet with different environmental conditions and, through their ability to vary, were more likely to survive if the natural surroundings altered in any way or if they were cut off from the rest of the group. Geographical isolation was still a necessary part of Darwin's scheme, but it was a noticeably different scheme. In 1856 he seemed to have a fully integrated and workable system, based on new findings yet retaining important concepts, like isolation, that could explain the production of new genera from old.

To weave everything together was now imperative for Darwin. Lyell had encouraged him to begin writing up his theories in the spring of 1856, and by the autumn of that year he had already finished two of the less complicated chapters—on variation under domestication and on geographical distribution.[32] But the most difficult chapters were yet to come. Would his ideas about the size of genera and the incidence of varieties hold up under the critical examination he gave to all his written work? More significant—since they were concerned with the very heart of his arguments—would his suggestions for the growth of genera and their fragmentation into smaller genera explain how groups of species increased in number, diverged from each other, and eventually died out as others took their place? Could he explain the history of life?

By the first months of 1857 Darwin had taken the plunge. In the sixth chapter of his manuscript *Natural Selection*, he tackled the vexed question of how forms gathered enough differences to turn into species, genera, and even families, as his theory required. He was convinced that selection could produce distinct lines of modification and was thus the cause of apparently "directional" evolution.

Without any formal explanation for divergence, Darwin did two things. He emphasised the role of competition, and he described the necessity for suitable "places" in the "polity of nature" for every step from varietal to specific rank. Competition for these places insured that only well-adapted variants succeeded in occupying them and that one form was always replaced by another even more closely adapted to the prevailing circumstances. This process of replacement appeared to move in or tend toward certain directions, a phenomenon which Darwin found difficult to explain. He fell back on the phrase "expression of variation in a right direction" to indicate—in an unconsciously teleological manner—such a movement.[33] It was an accurate if cumbersome phrase for the trends which he later described as divergence of character.

To begin his explanation Darwin emphasised that ecological "vacancies" were essential for evolution: "the existence . . . of an unoccupied or not perfectly occupied place is an all important element in the action of natural selection."[34] He went on to devote a longish section of Chapter VI to the possible ways in which niches could be vacated or made "not perfectly occupied" by their owners. Under the rubric "Causes favourable and unfavourable to Natural Selection," he worked his way through all the changes imaginable that could affect or increase the number of ecological nooks and crannies into which a modified variant could slip.

The second element of his argument was not so easy to describe, because Darwin thought that only "good" and "well-adapted" variants should be able to occupy such niches. "The expression of variation in a right direction implies that there is a place in the polity of nature, which could be better filled by one of the inhabitants after it has undergone some modification." The modification must be "in the right direction to profit the individual, otherwise it will not be selected"; it was pointless if the plant or animal simply fluctuated around a mean, as did polymorphic species, because these variations were "of no use in any one direction to the being in question" and did not allow particular characters to be "steadily selected, augmented and rendered nearly constant." Forthright though he was, this was special pleading with a vengeance. But it was the only way Darwin could account for one advantageous modification being added onto another and so start the ball rolling in "any given direction."[35]

Darwin finished his chapter on natural selection in March 1857. It focussed on the question of diverging lines of evolution—a "right direction"—without as yet any idea of a principle of divergence. However, it must be emphasised that the composition of Chapter VI did set the scene for Darwin's later work on divergence. Throughout the period of writing he was ever alive to the possibilities of clarifying the ideas and drawing in new correlations or more effective arguments. The "big species book" was an unparalleled opportunity for him to test his own ideas *in extenso* and to uncover flaws in his reasoning or evidence. When writing out some sixty or seventy pages on the question of natural selection and its apparently directional results, Darwin must have reflected deeply on what it was he was trying to say, whether it could be said more effectively, and—most of all—whether what he said truly described the manner in which he supposed natural selection to work. Darwin was brought to the point where his ideas, once so clear in his mind, were now clouded by doubts. He must have seen that "the expression of variation in a right direction" was ineffective, and even misleading, and must then have begun casting around for a solution. Darwin was poised for a major reformulation of his thoughts.

THE PRINCIPLE OF DIVERGENCE

The only contemporary evidence relating to Darwin's discovery of an explanation for divergent lines of evolution is to be found in his personal correspondence. He wrote to Hooker in August 1857, describing a few botanical calculations, and added a rider on the "principle of divergence" as it bore on his general theory: "If it will all hold good [his botanical arithmetic] it is very important for me; for it explains, as

I think, all classification, ie the quasi-branching and sub-branching of forms, as if from one root, big genera increasing and splitting up, etc as you will perceive. But then comes in, also, what I call a principle of divergence, which I think I can explain but which is too long, and perhaps you would not care to hear."[36]

Further corroboration appears in a letter written four weeks later to Asa Gray, in which Darwin summed up the whole theory of evolution by natural selection (the nucleus of his Linnean Society paper of July 1858)[37] and in which he included a closely argued paragraph on "one other principle" which "may be called the principle of divergence." He told Gray:

> The varying offspring of each species will try (only few will succeed) to seize on as many and as diverse places in the economy of nature as possible. Each new variety or species when formed will generally take the place of, and so exterminate its less well-fitted parent. This I believe to be the origin of the classification or arrangement of all organic beings at all times. These always *seem* to branch and sub-branch like a tree from a common trunk; the flourishing twigs destroying the less vigorous—the dead and lost branches rudely representing extinct genera and families.[38]

These letters to Hooker and Gray described divergence as it was ultimately to appear both in the *Origin* and in an addition to the *Natural Selection* manuscript composed in 1858. It was the most *diverse* offspring which managed to live and contribute successfully to the next generation: "The more diversified the descendents from any one species become in structure, constitution and habits, by so much will they be better enabled to seize on many and widely diversified places in the polity of nature, and so be enabled to increase in numbers."[39] In short, the most different variety in any bunch would be the one favoured by natural selection.

Obviously, therefore, Darwin had established this principle by the end of August 1857. Yet it had meant nothing to him the previous March, when he was putting the final touches to Chapter VI (see table). Between April and July he must have come across a flaw in his argument that led him to reassess the various elements of evolutionary theory. But what was the flaw?

On 14 July 1857—just three months after completing the first draft of Chapter VI—Darwin had discovered an elementary mistake in the calculation of his botanical arithmetic. The mistake was pointed

Table of Events in the Composition of *Natural Selection* and the Development of Darwin's "Principle of Divergence"

Year	*Date*	*Activity*	*Chapter*
1854	9 Sept.	"Finished packing up all my Cirripedes"*	
		"Began sorting notes for Species theory"	
	Nov.	Took up topic of aberrance and "creation"	
1855	Jan.	Began investigation into large genera	
1856	14 May	"Began by Lyell's advice writing Species Sketch"	
	13 Oct.	Finished whole of "Variation under Domestication"	II
		Finished first part of "Geographical Distribution"	XI
	16 Dec.	Finished whole of "Crossing"	III
1857	26 Jan.	Finished first part of "Variation under Nature"	IV
	3 Mar.	Finished whole of "Struggle for Existence"	V
	31 Mar.	Finished first draft "On Natural Selection"	VI
		Listed "Divergence" in Table of Contents	
	5 July	Finished whole of "Laws of Variation"	VII
	14 July	Letter to John Lubbock about mistake in his arithmetic	
		Began correction of botanical arithmetic	
	Aug.	Decided to repeat all calculations	
	22 Aug.	Letter to Hooker describing "Principle of Divergence"	
	5 Sept.	Letter to Asa Gray describing "Principle of Divergence"	
	29 Sept.	Finished whole of "Difficulties on the Theory"	VIII
	29 Dec.	Finished whole of "Hybridism"	IX
1858	9 Mar.	Finished whole of "Mental Powers and Instincts"	X
	Apr.	Finished revising botanical arithmetic	
	14 Apr.	Began writing "Discussion on large genera & small, & on Divergence and correcting chapter VI"	
	12 June	"Finished" above; added section on large genera to chapter IV and section on divergence to chapter VI	
	18 June	Interrupted by letter from A.R. Wallace	

*Quotations come from Darwin's *Journal*.

out by his young friend John Lubbock. "You have done me the greatest possible service in helping me to clarify my brains," wrote Darwin gratefully. "If I am as muzzy on all subjects as I am on proportion and chance—what a book I shall produce! . . . I am quite shocked to find how easily I am muddled, for I had before thought over the subject much and concluded my way was fair. It is dreadfully erroneous. What a disgraceful blunder you have saved me from. . . . But oh! if you knew how thankful I am to you!"[40]

Although Darwin openly encouraged Lubbock in all scientific matters and often indulged in innocent flattery, such uninhibited enthusiasm was quite unlike the even tenor of his normal correspondence with this friend and protégé, and perhaps mirrored a deep sense of shock on Darwin's part. For his remarks were a just assessment of the situation: Darwin had indeed been too simplistic in his protracted analysis of species, genera, and varieties. As if to spread the burden of this sudden and painful revelation, Darwin fired off a similar letter to Hooker on the same day, calling himself the "most miserable, bemuddled, stupid dog in all England," ready to "cry with vexation at my blindness and presumption." Historians can only be thankful that he did not carry out his extravagant threat to tear up his manuscripts and "give up in despair."[41]

In a nutshell, the reformulated method of computation required the entire Flora, or list of plants, to be divided into two groups according to the size of the genera *before* any calculations were carried out. Where Darwin had originally asked whether genera possessing the desired quality (such as varieties, or wide range, or whatever) also presented large numbers of species, he now looked at the problem the other way around and calculated whether some predetermined "large" genus had more of whatever was under discussion than correspondingly "small" genera.

In all previous computations he had divided his sample—a catalogue of plants—into two groups of genera, one of which contained the variable and the other of which did not. Occasionally he used a third category, the entire population. From these he derived the average number of species per genus for each category and then compared the results. Whichever group showed the higher average was considered to be the larger of the two. Such a method did not, strictly speaking, prove what Darwin thought it did. Even if genera with species that vary had, on the average, more species per genus than those without, it was still possible that varying species occurred most frequently in small (or middle-sized) genera.

Furthermore, although not categorically incorrect, Darwin's idea of "large" throughout his earlier statistics was, in reality, simply a statement of "bigger than." He seemed to have no concept of any absolute largeness or bigness, but merely compared one average with another and believed his point proved if one was larger than its pair. Since this was only a relative estimate of generic size, Darwin's results often showed undesirable fluctuations between one calculation and the next. Lubbock undoubtedly seized on these discrepancies and pointed out that Darwin was calculating relative largeness when his conclusions spoke of some real difference in size. Lubbock therefore abolished all connotations of relativity and substituted a division of the population into two halves, one containing all the truly large genera and one the small. Now, even if people quibbled with Darwin over his definition of large, at least *he* had defined it in unequivocal terms. The central question was still the one posed by Darwin in earlier computations: do varieties (or commonness or wide range) occur most frequently in the larger genera? But it was rephrased by Lubbock to ensure a more rigorous reply.[42]

Lubbock's changes forced Darwin to look not at the relative sizes of genera, but at the absolute "bigness" or "smallness" that each possessed. The emphasis shifted to the success that large genera so evidently enjoyed. Darwin suddenly realised that it was not just variation and the fortuitous production of "good" adaptations that induced large genera to produce yet more and more species, but also their potency. Large genera really were more successful than small. They were, in fact, the very acme of success, being more widespread and more abundant in individuals than their smaller confreres and also turning out more varieties among which more "good" adaptations were likely to emerge. Large genera were the winners, and their size was a definite statement about their superior position in life. In a biblical turn of phrase, Darwin asserted that "in the great scheme of nature, to that which has much, much will be given."

It was this notion of success and its corollary of winner takes all that allowed Darwin to fuse together ideas that had hitherto been separate in his mind. All at once things fell into place. As he later remembered in his autobiography, the "solution" suddenly occurred to him while in his carriage; he could still remember "the very spot in the road" and the feelings of joy and astonishment which accompanied it. Like "Colombus and his egg," the answer was startlingly obvious once it was discovered.[43]

Insofar as we can decide what may have been going on in anyone's mind, this reassortment of details can be reconstructed as follows. Prompted by Lubbock, Darwin recognised that large genera had more advantages than most. This was why they were more widespread and more numerous in individuals. Where before he had spoken only of forms being "better" adapted to their surroundings, now he had real advantages to deal with. The varieties produced in such numbers from the bigger genera should also be superior to ordinary variants, if his ideas about the inheritance of characters were to hold true. Moreover, since natural selection dictated that good variations were always preserved, what happened to this wealth of superior variants? Here, Darwin invoked the division of labour, a well-known idea[44] that permitted any number of individuals to coexist as long as they were more or less distinct from one another. In Darwin's eyes this meant that more forms of life could be supported in any one area, since they all performed different tasks and carried out different functions in the overall scheme of existence.

Confronted with a vision of many superior variants vying with each other for "places" in the economy of nature, and with the rule that only the most diversified set of individuals would manage to live together, Darwin progressed easily to the proposal that it was the most distinct or extreme variety that was favoured by natural selection. In a word, then, it was diversify or be done for. This single rule could be applied across the board and accounted for divergence among individual organisms, among varieties and species, and among all the higher taxonomic categories. The entire natural world displayed a tendency for increasing differentiation between its parts; along with the action of natural selection, there was evidently a "principle of divergence" constantly at work.

With the creation of this principle Darwin found a compelling reason for revising his botanical arithmetic. Here were striking new concepts which depended to a large degree on the conclusions derived from arithmetical statements about the appearance of varieties and the properties of large genera. Darwin realised that divergence could be seen most often—was most likely to occur—in the larger groups in nature, and particularly in the larger genera, which presented the most varieties and hence fuelled the process. Large genera were the ones that grew, split, and grew again, making up the major branches in the tree of life. But Lubbock had shown that his arithmetic was wrong. A cautious man, Darwin would not discuss the principle of divergence in print until he was confident of its statistical base; he decided to rework this extensive body of data, even though the "big book" was more than half completed and would have to be substantially revised. It is a measure of the significance of this constellation of new ideas that Darwin was prepared to commit himself to an interminable period of drudgery, to sacrifice whatever time it would take to go through an exercise that had already occupied most of his precious time for twenty months. "The subject," he told Hooker, "is in many ways so very important for me."[45]

It was not until the spring of 1858 that Darwin was perfectly satisfied with his arithmetical data, and then he returned immediately to the taxing question of divergence. Indeed, he noted in his journal that the weeks between 14 April and 12 June 1858 were devoted to a double writing-up: he drew up a "discussion on large genera and small" to be added to Chapter IV ("Variation under Nature") and at the same time wrote on "Divergence and correcting Chapter VI." The implication is that each account depended on the other. The interdependence of these two additions is emphasised by Darwin's concluding note on 12 June that both were "finished"—effectively lumping the two subjects together in practical terms, if not in intellectual ones.[46] The "principle of divergence" was at last a part of his account of evolution. Two days later, in a coincidence that still grips the imagination, Darwin received Wallace's unexpected letter from Ternate.

GEOGRAPHICAL DISTRIBUTION AND DIVERGENT EVOLUTION

Under this interpretation of Darwin's work before the *Origin*, the emergence of a principle of divergence can be seen as the last leg of a long inquiry into the general issue of divergent evolution. Between 1842 and 1858 Darwin approached the question from a number of angles: at times he thought the problem was solved; at others it ballooned out in a disturbing and temporarily uncontrollable fashion, forcing him to reevaluate previous arguments, to gather new information or reinterpret the old, and to provide reformulated explanations. It was a see-saw existence. Many of the phenomena of divergent evolution noted by Darwin through the years 1837–1840 found an explanation in the sketches of 1842 and 1844. Having dealt with them to the best of his ability, Darwin turned to a study of barnacles, no doubt to corroborate his writings in other ways. But the barnacles disclosed a whole range of evidence, obliging him to return to the thesis of 1844 in order to alter and expand various lines of argument. In the light of unlimited variation in nature Darwin undertook numerical studies of varieties, species, and genera to determine the source of new species. Over a span of many months (from 1855 to the middle of 1857) this botanical arithmetic indicated that large genera were more "fertile" than small. Never one to leave a fact unexplained or a question unasked, Darwin noted that if a fertile genus produces more and more species, these species will merely remain variations on a single theme unless divergence intervenes. How could the genus split into several genera? Before 1857 he answered this question with a somewhat hazily formulated scheme of geographical isolation, depending for the most part on results drawn from his arithmetical calculations on the wide geographical areas covered by species-rich and variable genera. Yet when he came to order these thoughts into a written account for the "big species book," the argument failed him. Divergence—the "expression of variation in a right direction"—still lacked an adequate explanation. While he was endlessly turning the problem over during the first six months of 1857, a relatively trivial event, not immediately concerned with divergence (although intimately connected with his numerical studies), caused Darwin to stop in his tracks. The reorganisation of his arithmetic stimulated a reorganisation of the issue of divergence. The various pieces of the puzzle were reassociated and reassembled in the middle of 1857, producing the much vaunted "principle." As an explanation it had much to commend it, and Darwin was eager to provide the necessary substantiation; he delayed the revision of the long manuscript until the arithmetical basis of the concept was fully examined and then hurriedly wrote up his ideas. The principle of divergence was playing a large part in Darwin's theory by early 1858.

If there is any message from this sequence of events, it is that Darwin's theories changed and evolved as he himself grew older and more mature, and that the *Essay* and *Natural Selection*—and, indeed, the *Origin of Species*—represent only his considered opinion on the problem of species at any given point in time. There is no good reason to believe that Darwin's ideas were static from the *Essay* onward or to reject the possibility that the meaning of certain key concepts changed and developed during the same years. Darwin seems to have tried no less than three explanations for the origin of species: first, a scheme based on geological dynamics, where divergence occurred as a normal consequence of isolation and natural selection; second, where he used the properties of large genera to explain speciation and diversification; and third, where he found it impossible to continue without some additional "force" in nature and so reformulated his theory around a new "principle of divergence."

Looking back over these different theories, it is interesting to see how many and varied were the applications of geographical distribution. Darwin had made full use of established principles, extending and adding to these wherever necessary, drawing in accessory material from associated fields and applying the resulting cluster of ideas to various aspects of his argument for evolution. To begin with, the erection or removal of geographic barriers, the isolation of populations or the opening up of new channels of communication, were prerequisites for alterations in the arrangement of living beings and for stimulating the production of variant offspring. Geographical seclusion was necessary to preserve any innovations that might arise and spread themselves through each small, insular population. Geographical communication was necessary to enable the distribution of these new forms as far as physical barriers and other obstacles would allow. So, in its own way, animal and plant distribution was as important for Darwin's first ideas about evolution as the concept of natural selection itself.

Continuing work on the dispersal of seeds and larvae by various accidental or occasional means of transport led Darwin to question the explanatory value of land bridges, and thence to the surrender of continental extensions as a prime agent in the dissemination of living beings. Having thus downgraded geological agents in one particular respect, Darwin went on to relinquish the geological foundations of his original 1842/1844 sketches of evolutionary change. He gradually abandoned the possibility of one grand scheme correlating the biological and geological kingdoms in a series of fluxes and dynamic movements, where organic change accompanied geological alterations and different topographic permutations activated great waves of plant and animal migration. That scheme had first been breached by his discovery of spontaneous, natural variation in barnacles and was successively eroded as he developed alternative explanations for every benchmark in the original argument. There was no second system large enough to embrace all Darwin's concepts as the tectonic scheme had done. The focus of his inquiries therefore shifted away from external agents such as geology to the many internal qualities of individual species of plants and animals.

Finally, taking the technique of botanical arithmetic from contemporary distributional studies, Darwin worked on the problem of what made one genus and its species more successful than another. These arithmetical surveys showed that distribution on a smaller geographical scale was crucial for a fuller understanding of organic change: the range of a species seemed to go hand-in-hand with the number of variations that could be expected within its ranks and was consequently linked in some way to the origin of new species or varieties. From such studies Darwin's principle of divergence emerged, a principle which appears to have played down the role of distribution in the evolutionary process. Once Darwin had an idea of divergent evolution he no longer depended on geographical isolation as an agent: divergence could take place without the intervention of topographic barriers, although Darwin rightly insisted that geographical isolation would accentuate such effects and encourage rapid differentiation. And, of course, organisms could disperse only when environmental conditions made it possible. But, at root, evolution with a principle of divergence was free of any logical requirement for isolation and geographic activity.

It may seem here that Darwin, the man who did so much to promote the study of animal and plant geography, actually played down distribution in his own writings. If true this would be ironic indeed. But Darwin did not devalue the subject of geographic distribution by allowing his new principle of divergence to bear the full brunt of the explanation; he merely shifted the emphasis of the argument. Instead of being an integral part of the theory of natural selection, geographical distribution now became the keystone of Darwin's demonstration of evolution in action. Two chapters on distribution in the *Origin of Species* took pride of place in his exposition of the many fields that may be explained by the hypothesis of organic change. Here, in the *Origin*, the intricate relations of distribution were utterly inexplicable by the independent creation of each species. But they were easily explained by colonisation from the most accessible source, together with the subsequent modification and better adaptation of the colonists to their new homes. Migration and modification were the keys to geographic arrangement, and, in turn, accounted for the various affinities between species of different countries. The organic world thus stood revealed as a product of the unceasing processes of distribution. The scope of Darwin's investigations, the breadth of his arguments, and the unifying nature of his conclusions in this area contributed substantially to the plausibility of the *Origin*'s thesis. Always an important subject for Darwin, geographical distribution turned out to be the most significant of all in establishing the likelihood of evolutionary change. Distribution studies had come a long way from Noah's Ark.

NINE

Conclusion

It is often assumed that the concepts of organic change first put forward by Darwin and Wallace marked the beginning of modern biogeography. Nicholas Polunin and Ronald Good are not alone in believing that the study of geographical distribution was both primitive and barren without the seminal idea of animal and plant transformations.[1] But this was not at all the case. From Noah's Ark to the advent of evolutionary theory, the study of animal and plant geography displayed a rich and complex history, a tapestry of colourful, closely woven theories, embroidered by many hands and critically assessed by contemporary enthusiasts. From Athanasius Kircher to Darwin and Wallace, a long succession of talented naturalists devoted the best of their ability to geographical distribution, developing a wide range of explanations and building up a strong body of practical techniques. Ideas, speculations, carefully composed theories, misunderstandings, and sheer incomprehension combined to produce an exceptionally fertile branch of natural historical researches. The *Origin of Species* (1859) and *Geographical Distribution of Animals* (1876) appear to have closed, rather than opened, an important chapter in distributional inquiries.

But whether Darwin's and Wallace's work was the beginning of one era or merely the end of another is perhaps irrelevant. What is much more interesting is the emergence of a specialised branch of natural history and the process by which scientific theories and techniques crystallised around a common core. The history of biogeography is more than simply a backcloth to evolutionary theory. In a small but not unimportant way, it shows how different branches of scientific thought can become separated from each other and sheds fresh light on some of the various factors influencing the development of a new field of inquiry.

Geographical distribution has long rested uneasily on the boundary between geography, geology, and natural history, pulled first one way and then another, always fluid and subject to a multitude of different interpretations. Since the middle of the eighteenth century, when any semblance of internal coherence sank with the idea of Noah's Ark, naturalists have been trying to fit topographical phenomena into a wider pattern of scientific explanation. The search has always been for an appropriate frame to fit the portrait.

By the second quarter of the nineteenth century, not one but two picture frames had been found. There were two distinct ways of looking at geographical distribution, one originating from the proto-social science of demography, the other from geology. For a time, the aims, methods, and *mores* of these divergent groups were radically different. Although the two were not necessarily incompatible, in practice an individual naturalist understood only one or the other. At a personal level, previous experience, natural preference, and the interests of close scientific colleagues must have helped determine which to adopt. It seems too that the naturalist's original discipline or subject area was also significant. Botanists and zoologists, on the whole, were more interested in the topography and comparative relations of groups of organisms than in the history of distribution. Geologists and palaeontologists, on the other hand, wanted mostly to expand their knowledge of the earth's past and to investigate organic arrangements as part of a larger process of inorganic change. Two different ways of "seeing" had emerged.[2]

The effect of two such distinct lines of distributional inquiry was not what might be expected. There was no confrontation and collision, no victory for one or the other interpretive style. Such martial language hardly reflects the nineteenth-century view of things, where puzzled incomprehension and shadowy half-understandings were more prevalent than academic cut and thrust. Friction—such as it was—between the two parties did not push them irrevocably apart but enabled them instead to create a common ground where both points of view were appreciated. Nor was this common ground mapped out beforehand. Rather, it emerged from the debates surrounding the problems of distribution that had most taxed early nineteenth-century thinkers—the questions of disjunct and geographically representative organisms. These provided the landmarks, as it were, for some larger meeting place of minds.

Neither issue had any ready-made explanation, for humanistic analogies could be just as fruitful as geological. The intricacy of the two problems under discussion and the lack of any central distributional dogma exacerbated uncertainty and encouraged naturalists to be broad-minded in the selection of the evidence they employed, tolerant of the various ways it was then put to use. Even though there were now different methods of interpreting nature, it was not clear that these were potential sources of conflict, possible rallying points for a clash of world-views. Rather, there was a miscellany of contemporary facts, theories, and assumptions reflecting, first and foremost, the preoccupations of the day. Distribution studies, in effect, were open to all comers.

The very flexibility of early-nineteenth-century distribution therefore created a forum for a wide variety of specialist natural historians. It brought together palaeontologists and systematists, botanists, zoologists, and comparative anatomists, the practical and theoretical, the gentleman amateur and the professional expert. They examined animal and plant geography from many different viewpoints, creating a scientific *embarras de richesses*. Concepts derived from one branch of the organic kingdom were liberally applied to others, and intellectual curiosity shifted from zoology to botany, demography to geology, shells to insects to plants, to fossils, and back again to existing species of both organic realms. Each shift was governed by changes not in distributional hypotheses themselves, but in the relevant theoretical context—in man's attitude toward the Bible, the history of life, the possibility of extinction. Each drew in further naturalists, perhaps with specific interests in, say, fossils, mammals, or ornithology; and each transformed the study of geography, stimulating new inquiries and more often than not investing different sections of the living world with novel, provocative meaning.

Few of these innovations were ever lost. Geographical distributionists tended to hold onto conceptual achievements, adding them to an already diverse body of abstract speculation to make yet another layer in a growing scaffold of ideas. Concepts derived from the study of plants were superimposed on those from zoogeography, causing naturalists to look again at the animal world and add further observations in a sequence of contributions that resembled nothing so much as an over-decorated Victorian town hall. With a turret here, and a curlicue there, geographical distribution was a medley of styles, perspective, and intention. It was the cerebral equivalent of Keble College, Oxford.

Among this mental medley were the ideas on which modern biogeography is based: the concept of dynamism, both ancient and modern, geological and biological; the belief in history, sequence, and the effect of time; and an appreciation of the statistics of populations, the numerical connections between different groups of organisms. Yet such notions were only partially interlocked before the time of Wallace and Darwin. The two evolutionists were unique in recognising that distribution theory depended on the amalgamation of diverse lines of inquiry, on the marriage between the concepts of geologists and topographic naturalists. They alone studied the natural world as a combination of both history and statistics, processes and patterns.

Darwin and Wallace were certainly unusual in combining these two lines of thought. But they owed their breadth of view to the ambivalence and confusion of contemporary distributional theories. The mutual problems of geology and topography held their attention in tandem. If the two approaches had been more polarised, if they had represented completely different philosophies, perhaps with additional social or institutional barriers, it would have been more difficult for either man to bridge the gap. As it was, they were spoilt for choice of materials to build an appropriate theory. And the context was such that both made the same choice.

Nor did this mingling of styles complete the story. Later, when Darwin came to rethink his ideas, it was again to distribution that he turned for his revised theory of transmutation. The same subject gave him material for at least two markedly different schemes over an interval of nearly twenty years. As an intellectual endeavour, the richness of geographical distribution was matched only by the ingenuity of Darwin's mind.

Like any branch of science, therefore, geographical distribution had its founding fathers, its crucial concepts, and dramatic moments. Yet distribution's resemblance to other scientific disciplines stops abruptly there. No central dogma united workers, no paradigm emerged; few magisterial syntheses punctuated its existence, and young Turks failed to rebel against the established order. Even today, there are no institutes, international societies, or chairs of biogeography, no corporate identity or central school of thought that might unite its many varied practitioners—although an important journal was established in 1974 partly to alleviate this very problem.

Indeed, without strong social or conceptual boundaries to set it apart, the study of geographical distribution retains close links with its early nineteenth-century forebears—geology, botany, and zoology. In this respect it cannot strictly be compared with other, more clearly defined areas of scientific interest. With genetics, for example, personal ambition and overpowering social ties were a major force in creating the new field of investigation; the scientific implications of Mendel's work were not fully pursued until a group of relatively lowly research workers, with little to lose, were prepared to challenge conventional ideas and institutions.[3] Nor can biogeography be directly compared with mainly intellectual specialities within natural history, such as mycology or ornithology, where the subject matter more or less dictates the approach to be adopted.[4] By contrast, geographical distribution is one of those perplexing subjects where many different areas of thought meet and intermingle, where a multitude of techniques, attitudes, and styles of work are employed, where it is possible to be a biogeographer and yet remain a botanist or zoologist. In the closing words of the 1981 International Biogeography Symposium, it is, and shows every sign of remaining, "the least well defined, derived discipline in contemporary biology."

Notes

CHAPTER ONE: DESCENT FROM ARARAT

1. R. Hooykaas, *Religion and the rise of modern science* (Edinburgh: Scottish Academic Press, 1972); C. E. Raven, *Natural Religion and Christian Theology* (Cambridge: Cambridge University Press, 1953), vol. 1; R. S. Westfall, *Science and Religion in Seventeenth-Century England* (New Haven: Yale University Press, 1958); C. Hill, *Society and Puritanism in Pre-Revolutionary England* (London: Secker and Warburg, 1964); idem, *The Intellectual Origins of the English Revolution* (Oxford: Oxford University Press, 1965); idem, *The Century of Revolution, 1603–1714* (London: Nelson, 1980); C. A. Russell, ed., *Science and Religious Belief: A selection of recent historical studies* (London: University of London Press, 1973), including R. K. Merton, "Puritanism, Pietism and Science," reprinted from *Sociological Review*, o.s. 28, pt. 1 (1936).

2. P. Conor Reilly, *Athanasius Kircher, S.J.: Master of a Hundred Arts, 1602–1680* (Wiesbaden: Edizioni del Mondo, 1974).

3. Athanasius Kircher, *Arca Noë in tres libros digesta* (Amsterdam: Waesberg, 1675), and summarised by Don Cameron Allen, *The Legend of Noah: Renaissance Rationalism in Art, Science and Letters* (Urbana: Illinois University Press, 1949), pp. 182–191.

4. Allen, *Legend of Noah*, pp. 66–91.

5. Allen, *Legend of Noah*, pp. 155–73, and illustration following p. 176.

6. The frontispiece to G. Piso and G. Marcgravi, *Historia Naturalis Brasiliae* (Amsterdam: Elzevier, 1648), shows a sloth, lemur, anteater, and pineapple, among other things. The second edition of 1658 depicts a dodo, rhinoceros, and porcupine. Further information from Wilma George, "Sources and background to discoveries of new animals in the 16th and 17th centuries," *History of Science* 18 (1980): 79–104.

7. Robert Hooke, *Lectures and Discourses of Earthquakes and Subterranean Eruptions*, in *The Posthumous Works of Robert Hooke* (London: R. Waller, 1705).

8. Richard S. R. Fitter, *The Ark in our Midst: The story of the introduced animals of Britain* (London: Collins, 1959), pp. 107, 114, 277.

9. Edward Brerewood, *Enquiries touching the diversity of languages and religions through the chief parts of the world.* (London: J. Bill, 1614).

10. Hugo Grotius, *De origine gentium Americanarum* (Paris and Amsterdam: n.p., 1642), translated by E. M. Goldsmid as *On the origin of the native races of America*, in his *Bibliotheca Curiosa* (Edinburgh: privately printed, 1883–1888).

11. Johannes Laetius, *Notae ad dissertationem H. Grotii de origine gentium Americanarum* (Paris and Amsterdam: n.p., 1643). The controversy between Grotius and Laetius continued, with Grotius issuing a "dissertatio altera" in 1643, followed by a rejoinder by Laetius in 1644.

12. Georgius Hornius, *De originibus Americanis libri quatuor* (Leyden: n.p., 1652), and Allen, *Legend of Noah*, pp. 113–37.

13. José de Acosta, *Historia natural y moral de las Indias* (Seville: J. de Lyon, 1590).

14. Justus Lipsius, *Physiologiae stoicorum, libri III*, in his *Opera* (Antwerp: n.p., 1637).

15. Georgius Hornius, *Arca Noae sive historia imperiorum et regnorum a condito orbe ad nostra tempora* (Leyden: n.p., 1666).

16. Matthew Hale, *The Primitive Origination of Mankind considered and examined according to the light of Nature* (London: W. Shrowsbery, 1677), pp. 197–203.

17. Isaac de La Peyrère, *Praeadamitae . . .* (London: n.p., 1655), translated as *Men before Adam, or a discourse upon the 12th, 13th and 14th verses of the Fifth Chapter of the Epistle of the Apostle Paul to the Romans* (London: n.p., 1656).

18. Abraham van der Myl [Mijl], *De origine animalium et migratione populorum* (Geneva: n.p., 1667).

19. Hale's criticisms in *Primitive Origination*, pp. 184–89.

20. Hale, *Primitive Origination*, pp. 203–05.

21. Edward Stillingfleet, *Origines Sacrae, or a rational account of the grounds of Christian faith* (London: H. Mortlock, 1662).

22. Matthew Poole, *Synopsis Criticorum aliorumque Sacrae Scripturae interpretum* (London: n.p., 1669–1676).

23. Jean Le Clerc, *Commentarii philologici et paraphrases in vet Testam* (Amsterdam: n.p., 1690–1731).

24. See for example, Allen, *Legend of Noah*, pp. 92–112; M. J. S. Rudwick, *The Meaning of Fossils: Episodes in the History of Palaeontology* (London: MacDonald, 1972), pp. 72–93; and Roy Porter, *The Making of Geology: Earth Science in Britain, 1660–1815* (Cambridge: Cambridge University Press, 1972), pp. 62–78.

25. John Woodward, *An Essay toward a natural history of the Earth . . . with an account of the universal Deluge* (London: n.p., 1695).

26. John Ray, *Miscellaneous Discourses concerning the dissolution and changes of the world* (London: S. Smith, 1692), pp. 66–98.

27. William Whiston, *A New Theory of the Earth* (London: n.p., 1696).

28. David Elliston Allen, *The Naturalist in Britain: A Social History* (Harmondsworth: Penguin Books, 1978), pp. 26–72.

29. C. Linnaeus, *Museum S.R.M. Adolphi Friderici Regis* (Holmiae: P. Momma, 1754) and translated by J. E. Smith, *Reflections on the study of Nature* (London: George Nicol, 1786), p. 14. See also Wolf Lepenies, "Linnaeus's *Nemesis divina* and the concept of divine retaliation," *Isis* 266 (1982): 11–27, which explains Linnaeus's theology in more detail.

30. C. Linnaeus, *Oratio de Telluris habitabilis incremento . . .* (Leyden: C. Haak, 1744) and translated by F. J. Brand in *Select dissertations from the Amonitates Academicae: A Supplement to Mr Stillingfleet's Tracts relating to Natural History* (London: G. Robinson, 1781), vol. 2, pp. 71–127.

31. Smith, *Reflections*, p. 18.

32. Brand, *Select dissertations*, vol. 2, pp. 81–86.

33. Michel Foucault, *The Order of Things: An Archaeology of the Human Sciences* (London: Tavistock, 1970); C. E. Raven, *English Naturalists from Neckham to Ray: A Study of the Making of the Modern World* (Cambridge: Cambridge University Press, 1947); and A. G. Morton, *History of Botanical Science* (London: Academic Press, 1981), contain some of the most helpful and interesting discussions of classification systems and the meanings that they represented.

34. C. Linnaeus, *Systema Naturae 1735, facsimile edition with a translation of the Observationes*, ed. M. S. J. Engel-Ledeboer and H. Engel (Nieuwkoop: n.p., 1964).

35. Smith, *Reflections*, p. 3.

36. F. N. Egerton, "Ancient Sources for Animal Demography," *Isis* 59 (1968): 175–89, and "A Bibliographical guide to the history of general Ecology and Population Ecology," *History of Science* 15 (1977): 189–215.

37. G. L. L. Buffon, *Barr's Buffon: Buffon's Natural History containing a theory of the earth* (London: J. S. Barr, 1792), vol. 1, p. 33.

38. *Barr's Buffon*, vol. 1, pp. 38–39.

39. *Barr's Buffon*, vol. 1, p. 12.

40. See also Gareth Nelson, "From Candolle to Croizat: Comments on the history of biogeography," *Journal of the History of Biology* 11 (1978): 269–305.

41. E. A. W. Zimmermann, *Specimen Zoologiae Geographicae Quadrupedum domicilia et migrationes sitens* (Leyden: T. Haak, 1777). See also F. S. Bodenheimer, "Zimmermann's *Specimen Geographicae Quadrupedum*, a remarkable zoogeographical publication of the end of the 18th century," *Archives Internationale d'Histoire des Sciences* 8 (1955): 351–57.

42. Gilbert White, *The Natural History and Antiquities of Selborne: The standard edition by E. T. Bennett*, rev. ed. J. E. Harting (London: Swan Sonnenschein, 1889), pp. 169, 249.

CHAPTER TWO: THE NATIONS OF PLANTS

1. Johann Reinhold Forster, *Observations made during a voyage round the world, on physical geography, natural history, and ethnic philosophy* (London: n.p., 1778), p. 174.

2. J. R. Forster, *Observations*, p. 169.

3. J. R. Forster, *Observations*, pp. 184, 185–186.

4. Described in *Herbarium Willdenow: Alphabetical Index*, ed. Paul Hiepko (Switzerland: Interdocumentation Company, 1972).

5. Karl Willdenow, *Principles of Botany* (Edinburgh: Blackwood, 1805), p. 371.

6. Willdenow, *Principles*, p. 373.

7. E. S. Russell, *Form and Function: A contribution to the History of Animal Morphology* (London: Murray, 1916), pp. 44–112; J. T. Merz, *A History of European Scientific Thought in the Nineteenth Century* (New York: Dover, 1965); and T. Lenoir, "The Göttingen School and the development of transcendental Naturphilosophie in the Romantic era," *Studies in History of Biology* 5 (1981): 111–205.

8. J. Théodoridès, "Humboldt and England (1769–1859)," *British Journal for the History of Science* 3 (1966): 39–55; and Adolf Meyer-Abich, *Alexander von Humboldt* (Bonn: Internationes [The Alexander von Humboldt Foundation], 1969).

9. H. Fieldler, S. Scheibe, E. Germer, *Georg Forster, Naturforscher, Weltreisender, Humanist und Revolutionar* (Wörlitz: Staatliche Schlosser, 1975).

10. Alexander von Humboldt, *Cosmos: A sketch of a physical description of the Universe*, trans. E. C. Otté (London: Bohn, 1849–1858), p. 357. See also his *Aspects of Nature in different lands and different climates with scientific elucidations*, trans. Mrs Sabine (London: Longman, 1849), vol. 2, p. 100.

11. Alexander von Humboldt, *Essai sur la géographie des plantes; accompagne d'un tableau physique des régions equinoxiales* (Paris: Levrault, 1807), and facsimile edition of the first part only by the Society for the Bibliography of Natural History, Sherborn Fund Facsimiles 1 (1959), p. 31. Since Stearn's analysis of the *Essai*'s publishing history it is usual to refer to this book as having been issued in 1807, not 1805 as is sometimes printed on the title page. See W. T. Stearn, "Humboldt's *Essai sur la Geographie des Plantes*," *Journal of the Society for the Bibliography of Natural History* 3, part 7 (1960): 351–57.

12. Humboldt, *Essai*, pp. 41–42. Loosely translated: "vegetation; animal life; geological structure; human occupation; air temperature; the lower limits of perpetual snow; the chemical composition of the atmosphere and its electrical resistance; barometric pressure; the decreasing effects of gravity; the intensity of the blue colour of the sky; the diminishing power of light rays and their refraction; and the temperature at which water comes to the boil at various heights."

13. Humboldt, *Aspects*, vol. 2, p. 15.

14. Alexander von Humboldt, *Personal Narrative of travels to the Equinoctal regions of the New Continent, 1799–1804*, trans. Helen Maria Williams (London: Longman, 1821–1825), vol. 1, p. xvi.

15. Harald Westergaard, *Contributions to the History of Statistics* (The Hague and Paris: Mouton, 1969); and Karl Pearson, *The History of Statistics in the 17th and 18th centuries against the changing background of intellectual, scientific and religious thought: Lectures given by Karl Pearson at Uni-*

versity College during the academic sessions 1921–1933, ed. E. S. Pearson (London: Griffin, 1978).

16. G. Achenwall, *Staatsverfassung der heutigen vornehmsten Europäischen Reiche*, 5th ed. (Göttingen: n.p., 1768).

17. Paul R. Sweet, *Wilhelm von Humboldt: A Biography* (Columbus: Ohio State University Press, 1978), vol. 1, pp. 123–200; and Lenoir, "The Göttingen School," pp. 115–28.

18. Quoted in Sweet, *Wilhelm von Humboldt*, vol. 1, pp. 138–39.

19. Maurice Crosland, *The Society of Arcueil: A view of French science at the time of Napoleon I* (London: Heinemann, 1967).

20. Augustin de Candolle, "Géographie botanique" in *Dictionnaire des sciences naturelles*, ed. F. C. Levrault (Paris: Levrault, 1820), vol. 18, p. 384.

21. Augustin de Candolle, "Mémoire sur la géographie des plantes de France, considerée dans ses rapports avec la hauteur absolue," *Soc. d'Arcueil Mem. de Phys.* 3 (1817): 262–322.

22. Augustin de Candolle, *Projet d'un flore physico-géographique de la vallée du Leman* (Geneva: Paschoud, 1821).

23. Augustin de Candolle, "Géographie botanique," p. 392. Loosely translated: "All or nearly all plants, left to themselves, tend to occupy a particular place on the earth; the determination of the laws that regulate these patterns of vegetation is the [aim of the] study of habitations."

24. Augustin de Candolle, *Theorie élementaire de la botanique*, 2nd ed. (Paris: Roret, 1819), p. 462.

25. Alexander von Humboldt, *Florae Fribergensis specimen* (Berlin: Rottman, 1793).

CHAPTER THREE: A SCIENCE OF PATTERNS

1. Nathanial Winch, *An Essay on the geographical distribution of plants* (Newcastle: Walker, 1819).

2. Alexander von Humboldt, "Géographie des plantes," in Augustin de Candolle, "Géographie botanique," in *Dictionnaire des sciences naturelles*, ed. F. C. Levrault (Paris: Levrault, 1820), vol. 18, pp. 422–32.

3. Alexander von Humboldt, *Prolegomena. Nova Genera et Species Plantarum quas in peregrinatione ad plagam eaquinoctiatem Orbis Novi collegerunt, descripserunt et adumbraterunt . . .* , by A. Bonpland, A. von Humboldt, and C. S. Kunth (Paris: Lutetiae, 1815–1825), p. viii.

4. Humboldt, *Prolegomena*, p. xiv.

5. Humboldt, *Prolegomena*, p. xviii.

6. Alexander von Humboldt, *Aspects of Nature in different lands and different climates with scientific elucidations*, trans. Mrs Sabine (London: Longman, 1849), vol. 2, pp. 108, 111.

7. Robert Brown, *General Remarks, geographical and systematical, on the botany of Terra Australis* (London: n.p., 1814), p. 5.

8. Brown, *General Remarks*, pp. 35–36, 55.

9. Augustin de Candolle, "Géographie botanique," p. 400. Loosely

translated: "To conclude [our discussion of] what is relevant to this type of botanical arithmetic, as M. de Humboldt calls it, and to show the extent to which it can describe the general picture of the vegetation of different countries, I repeat that one must focus on one aspect of the comparison of the proportional number of species and genera of a country. The more limited the average number of species in each genus or each family, the more the appearance of the vegetation shows variety; on the contrary, the larger that number the more a glance at the countryside shows monotony of form."

10. P. Jaccard, "Sur le cóefficient generique," *Chronica Botanica* 6 (1941): 361–64, 389–91.

11. Leopold von Buch, *Description physique des iles Canaries, suivie d'une indication des principaux volcans du globe*, trans. C. Boulanger (Paris: n.p., 1836), pp. 145–50.

12. Darwin Archive, Cambridge University Library Archives, Dar 121, pp. 156–58.

13. Rudolf Wagner, "The numerical relations of animals from Linnaeus to the present day," *Edinburgh New Philosophical Journal* 15 (1833): 221–27.

14. H.C. Watson, "Observations made during the summer of 1832 on the temperature and vegetation of the Scottish Highland mountains," *Edinburgh New Philosophical Journal* 14 (1833): 323.

15. H.C. Watson, *Cybele Britannica: or British plants and their geographical relations* (London: Longman, 1847–1859, Supplement 1860, Compendium 1868–1870), vol. 1, p. 42.

16. Watson, *Cybele*, vol. 4, p. 379.

17. F. J. F. Meyen, *Outlines of the geography of plants with particular enquiries concerning . . . the principal cultivated plants on which the prosperity of nations is based*, trans. Margaret Johnson (London: Ray Society, 1846), pp. 260, 270.

18. Humboldt, *Aspects*, vol. 2, p. 113.

19. C. G. Ehrenberg, *Die Infusionsthierchen als volkommene Organismen* (Leipsig: Voss, 1838).

20. Karl Illiger, "Tabellarische Ubersicht uber die Vertheilung der Vogel uber die Erde," *Abhandlungen der Königlichen Akademie der Wissenschaften zu Berlin* (1812–1813): 221–36.

21. P. André Latreille, "Introduction a la géographie générale des Arachnides et des Insectes," *Edinburgh New Philosophical Journal* 5 (1821): 370–78, and ibid. 6 (1822): 51–62. See pp. 373, 377–78, 58–59.

22. Théodore Lacordaire, "On the geographical distribution of insects," *Edinburgh New Philosophical Journal* 27 (1838–1839): 94–111, 333–51, and ibid. 28 (1839–1840): 170–83, 224–38.

23. Lacordaire, "Geographical distribution," 27: 346.

24. Edward Forbes, "Report on the Mollusca and Radiata of the Aegean Sea, and on their distribution, considered as bearing on geology," *Reports of the British Association for the Advancement of Science* 1843 (1844): 130–93.

25. James Dana, "Geographical distribution of Zoophytes," *Edinburgh New Philosophical Journal* 43 (1847): 41–45.

26. Watson, *Cybele*, vol. 4, p. 441.

27. Edward Forbes, "On the connexion between the distribution of the existing fauna and flora of the British Isles, and the geological changes which have affected their area, especially during the epoch of the northern drift," *Memoirs of the Geological Survey of Great Britain* 1 (1846): 336–432; and correspondence between J. D. Hooker and H. C. Watson, Archives, Royal Botanic Gardens, Kew, J. D. Hooker incoming letters, vol. 8, fol. 159, 163–64.

28. R. D. Altick, *Victorian People and Ideas* (New York: Norton, 1973); and M. J. Cullen, *The Statistical Movement in early Victorian Britain* (Sussex: Harvester Press, 1975).

29. E. S. Pearson and M. G. Kendall, eds., *Studies in the History of Statistics and Probability* (London: Charles Griffin, 1970).

30. D. V. Glass, "Some aspects of the development of demography," *Journal of the Royal Society of Arts* 104 (1956): 854–69, and *Numbering the People: The eighteenth-century population controversy and the development of census and vital statistics in Britain* (London: D. C. Heath, 1973).

31. S. F. Cannon, *Science in Culture: The early Victorian period* (London and New York: Dawson, 1978), pp. 225–62.

32. Watson, *Cybele*, vol. 1, p. 12.

33. Charles Lyell, *Principles of Geology, being an attempt to explain the former changes of the earth's surface by reference to causes now in operation* (London: Murray, 1830–1833), vol. 2, pp. 31–32. Martin Rudwick has meticulously examined Lyell's use of human, demographic, and historical analogies: M. J. S. Rudwick, "Historical analogies in the geological work of Charles Lyell," *Janus* 64 (1977): 89–108; idem, "Transposed concepts from the human sciences in the early work of Charles Lyell," in *Images of the Earth: Essays in the History of the Environmental Sciences*, ed. L. J. Jordanova and Roy Porter (Chalfont St. Giles, Bucks.: British Society for the History of Science Monographs, 1979); and idem, "Charles Lyell's dream of a statistical palaeontology," *Palaeontology* 21 (1978): 225–44.

34. Cannon, *Science in Culture*, pp. 81, 116.

35. John Herschel, *Preliminary Discourse on the study of Natural Philosophy* (London: Lardner's Cabinet Cyclopedia, 1830).

36. Watson, *Cybele*, vol. 1, title page.

37. Darwin Archive, Cambridge University Library Archives, Dar 104, p. 219.

38. Charles Darwin, *More Letters of Charles Darwin, a record of his work in a series of hitherto unpublished letters*, ed. Francis Darwin and A. C. Seward (London: Murray, 1903), vol. 1, p. 403.

39. J. D. Hooker and Thomas Thomson, *Flora Indica. Volume I with an Introductory Essay*, 1 vol. (London: W. Pamplin, 1855), pp. 10–11.

40. J. D. Hooker, *Flora Novae Zelandiae* (London: Reeve Brothers, 1853–1855), part of Hooker's *Botany of the Antarctic Voyage of H.M.S. Erebus and Terror . . . 1839–1843* (London: Reeve Brothers, 1844–1860), p. xxvi.

41. Hooker and Thomson, *Flora Indica*, p. 40.

42. Alphonse de Candolle, *Géographie botanique raisonnée, ou exposition des faits principaux et des lois concernant la distribution géographique*

des plantes de l'époque actuelle (Paris: Masson, 1855), p. xvi: "There is in every science, art, or field of research, a part that is statistical or numerical. One finds it in agriculture, in medicine, in all branches of administration, in the physical, natural, and even the moral sciences. Statistics are important in botanical geography. For my own part, I admit, I love figures where others detest them."

43. Alphonse de Candolle, *Géographie botanique raisonnée*, p. xvii. "This appreciation of methodology as well as the discussion of facts and the tendency to investigate the causes of phenomena, has seemed to me to establish a rather large difference between my work and the majority of treatises on botanical geography, in which the authors are content to describe the countryside or to recite the facts."

44. M. J. S. Rudwick, "The strategy of Lyell's *Principles of Geology*," *Isis* 61 (1970): 5–33; and R. S. Porter, "Charles Lyell and the principles of the history of geology," *British Journal for the History of Science* 9 (1976): 91–103.

CHAPTER FOUR: A SCIENCE OF PROCESSES

1. Peter J. Bowler, *Fossils and Progress: Paleontology and the idea of a progressive evolution in the nineteenth century* (New York: Science History Publications, 1976).

2. Adam Sedgwick, "Anniversary Address," *Proceedings of the Geological Society of London* 1 (1831): 295.

3. Baron Georges Cuvier, *Recherches sur les ossemens fossiles de quadrupèdes, où l'on rétablit les caractères de plusieurs espèces d'animaux que les révolutions du globe paroissent avoir détruites* (Paris: Chez Deterville, 1812). Quoted from William Coleman, *Georges Cuvier, Zoologist: A Study in the History of Evolutionary Theory* (Cambridge, Mass.: Harvard University Press, 1964), p. 119.

4. Georges Cuvier and Alexandre Brongniart, *Essai sur la géographie minéralogique des environs de Paris* (Paris: Baudouin, 1811), included in Cuvier's *Recherches*, as above.

5. James Hutton, *Theory of the Earth, with proofs and illustrations* (Edinburgh: Cadell and Davies, 1795); and P. A. Gerstner, "The reaction to James Hutton's use of heat as a geological agent," *British Journal for the History of Science* 5 (1971): 353–362.

6. J. B. Fourier, "Mémoire sur le refroidissement seculaire du globe terrestre," *Bulletin de la Société Philomathique de Paris* (1820): 58–70.

7. E. F. von Schlotheim, *Beschreibung merkwurdiger Krauter-Abdrücke und Pflanzen-Versteinerungen* (Gotha: Becker, 1804), and "Abhandlung uber die Krauter-Abdrücke im Schieferthon und sandstein der Steinkohlen-Formationen," *Hoffmagazin* 1 (1801): 76–95.

8. Adolphe Brongniart, "General considerations on the nature of the vegetation which covered the surface of the earth, at the different epochs of the formation of its crust," *Edinburgh New Philosophical Journal* 6 (1828–

1829): 350, and "On the great divisions of the vegetable kingdom occurring in the different geological formations," ibid. 40 (1846): 285. By 1846 fossil species had nearly quadrupled in number to a total of 1792, while the known species of living plants had increased to roughly 80,000. Fossils therefore constituted about one forty-fifth of known species. For a general history of palaeobotany see H. N. Andrews, *The Fossil Hunters: In search of Ancient Plants* (Ithaca: Cornell University Press, 1980), especially chap. 4; and Y. Conry, *Correspondence entre Charles Darwin et Gaston de Saporta* (Paris: Galien, 1972), introduction.

9. Adolphe Brongniart, *Prodrome d'une histoire des végétaux fossiles* (Paris: Levrault, 1828), pp. 177–78.

10. Adolphe Brongniart, *Prodrome*, p. 220.

11. Alphonse de Candolle, "On the history of fossil vegetables," *Edinburgh New Philosophical Journal* 18 (1834): 85, 98–99.

12. Alphonse de Candolle, "Fossil vegetables," p. 97.

13. Alphonse de Candolle, "Fossil vegetables," p. 87.

14. Alphonse de Candolle, "Fossil vegetables," p. 94.

15. John Fleming, "On the value of the evidence from the animal kingdom, tending to prove that the Arctic regions formerly enjoyed a milder climate than at present," *Edinburgh New Philosophical Journal* 6 (1829): 277–86.

16. Alexander Murray, "Thoughts regarding the influence of rock on native vegetables," *Edinburgh New Philosophical Journal* 11 (1831): 63.

17. Anonymous, "A uniformity of climate prevailed over the earth prior to the time of the Deluge," *Edinburgh New Philosophical Journal* 8 (1830): 366–67.

18. Georges Cuvier, "Mémoire sur les espèces d'elephans vivantes et fossiles," *Mémoires de l'Institut National des Sciences et Arts (Classe Mathématiques et Physiques)* 2 (1799): 1–22.

19. J. Pentland, "On the fossil bones of Wellington valley, New Holland or New South Wales," *Edinburgh New Philosophical Journal* 12 (1832): 301–08. Joseph Pentland was mistakenly called William Pentland in several journals, including the Royal Society Catalogue of Scientific Papers. See W. A. S. Sarjeant and J. B. Delair, "An Irish naturalist in Cuvier's laboratory: The letters of Joseph Pentland, 1820–1832," *Bulletin of the British Museum (Natural History)*, Historical ser. 6, no. 7 (1980): 245–319.

20. Robert Jameson, "On the fossil flora," *Edinburgh New Philosophical Journal* 13 (1832): 349–50.

21. John Lindley, "On the mode of determining fossil plants," *Edinburgh New Philosophical Journal* 13 (1832): 221–28.

22. Communicated to Jameson in 1831 and printed in the *Edinburgh New Philosophical Journal* 10 (1831): 364–71.

23. Announced in "Scientific Intelligence," *Edinburgh New Philosophical Journal* 10 (1831): 383.

24. W. Clift and R. Jameson, "Report by Mr Clift in regard to the fossil bones found in the Bone Caves and Bone-Breccia of New Holland," *Edinburgh New Philosophical Journal* 10 (1831): 395.

25. Pentland, "On the fossil bones," p. 308.

26. W. Clift and Woodbine Parish, "An account of the discovery of portions of three skeletons of the Megatherium in the province of Buenos Ayres," *Proceedings of the Geological Society of London* 1 (1832): 403–04.

27. E. W. Brayley, "On the odour exhaled from certain organic remains in the Diluvium of the Arctic circle," *Philosophical Magazine* 9 (1831): 411–18.

28. Charles Darwin, *The Voyage of the Beagle*, intro. H. G. Cannon (London: Dent, Everyman, 1959), pp. 77–80, 123–25.

29. Charles Darwin, "A sketch of the deposits containing extinct mammalia in the neighbourhood of the Plata," *Proceedings of the Geological Society of London* 2 (1837): 542–44.

30. Richard Owen, *On the Nature of Limbs* (London: J. van Voorst, 1849).

31. Adrian Desmond, "Designing the Dinosaur: Richard Owen's response to Robert Edmond Grant," *Isis* 70 (1979): 224–34.

32. Louis Agassiz, "The primitive diversity and number of animals in geological times," *Edinburgh New Philosophical Journal* 57 (1854): 286–87.

33. Alcide d'Orbigny, "General considerations regarding the palaeontology of S. America compared with that of Europe," *Edinburgh New Philosophical Journal* 35 (1843): 369, 372.

34. Roderick Impey Murchison, *The geology of Russia in Europe and the Ural mountains* (London: Murray, 1845), vol. 1, p. 19.

35. C. Lyell, *Principles of Geology, being an attempt to explain the former changes of the earth's surface by reference to causes now in operation* (London: Murray, 1830–1833), vol. 2, pp. 169, 173.

36. Lyell, *Principles*, vol. 1; and Dov Ospovat, "Lyell's theory of Climate," *Journal of the History of Biology* 10 (1978): 317–39.

37. Lyell, *Principles*, vol. 2, pp. 141–61, 176–77.

38. William Whewell, *The Philosophy of the Inductive Sciences, founded upon their history*, 2nd ed. (London: Parker, 1847), pp. 645–46.

39. A. Sedgwick, "Anniversary Address," *Proceedings of the Geological Society of London* 1 (1830): 211.

40. J. C. Prichard, "On the relations of Ethnology to other branches of knowledge," *Edinburgh New Philosophical Journal* 43 (1847): 308–309.

41. Whewell's "palaetiological" sciences characterised in William Whewell, *History of the Inductive Sciences, from the earliest to the present times* (London: Parker, 1837).

CHAPTER FIVE: THE EARTH AS A CHESSBOARD

1. Gareth Nelson, "From Candolle to Croizat: Comments on the history of biogeography," *Journal of the History of Biology* 11 (1978): 269–305.

2. Augustin de Candolle, "Géographie botanique," in *Dictionnaire des sciences naturelles*, ed. F. C. Levrault (Paris: Levrault, 1820) vol. 18, p. 400.

3. Théodore Lacordaire, "On the geographical distribution of insects," *Edinburgh New Philosophical Journal* 28 (1839–1840): 173.

4. K. P. J. Sprengel and Augustin de Candolle, *Elements of the Philosophy of Plants containing the Principles of Scientific Botany* (Edinburgh: W. Blackwood, 1821), p. 283.

5. Thomas H. Huxley, "Memoir of Professor Edward Forbes, F.R.S.," *Literary Gazette* 1975 (25 November 1854): 1017.

6. Edward Forbes, "Report on the Mollusca and Radiata of the Aegean Sea, and on their distribution, considered as bearing on Geology," *Reports of the British Association for the Advancement of Science* 1843 (1844): 171.

7. Forbes, "Mollusca and Radiata," p. 175.

8. E. Forbes, "On the light thrown on geology by submarine researches," *Edinburgh New Philosophical Journal* 36 (1844): 318–27, praised by Leonard Horner in "Anniversary Address of the President," *Quarterly Journal of the Geological Society* 2 (1846): 145–221, especially pp. 83–84, 161.

9. E. Forbes, "On the connexion between the distribution of the existing fauna and flora of the British Isles, and the geological changes which have affected their area, especially during the epoch of the northern drift," *Memoirs of the Geological Survey of Great Britain* 1 (1846): 336–432.

10. E. Forbes, "The life of an individual and the duration of a species," *Edinburgh New Philosophical Journal* 53 (1852): 130–35; and Charles Lyell, *Principles of Geology* (London: Murray, 1830–1833) vol. 2, pp. 128–41.

11. E. V. Wulff, *An introduction to historical plant geography . . .*, trans. E. Brissenden (Waltham, Mass.: Chronica Botanica Co., 1943), p. 202.

12. Charles Darwin, *The Life and Letters of Charles Darwin, including an Autobiographical chapter*, ed. Francis Darwin (London: Murray, 1887), vol. 2, pp. 38–39.

13. P. A. Latreille, "Introduction a la géographie générale des Arachnides et des Insectes," *Edinburgh New Philosophical Journal* 5 (1821): 371.

14. H. C. Watson, "Observations made during the summer of 1832 on the temperature and vegetation of the Scottish Highland mountains," *Edinburgh New Philosophical Journal* 14 (1833): 317–23.

15. M. J. S. Rudwick, "The glacial theory," *History of Science* 8 (1970): 137–57.

16. Louis Agassiz, *Studies on Glaciers*, trans. Albert V. Carozzi (New York: Hafner, 1967).

17. Bert Hanson, "The early history of glacial theory in British geology," *Journal of Glaciology* 9 (1970): 135–41; and F. J. North, "Centenary of the glacial theory," *Proceedings of the Geologists' Association* 54 (1943): 1–28.

18. Forbes, "On the connexion," p. 345.

19. R. I. Murchison, "Presidential address," *Reports of the British Association for the Advancement of Science* 1846 (1847): xxxiii–xxxiv.

20. Leonard Horner, "Anniversary Address of the President," *Quarterly Journal of the Geological Society* 3 (1847): xxxviii, lv–lxviii.

21. L. Horner, "Anniversary Address," p. lxvii; and Charles Darwin, *More Letters of Charles Darwin, a record of his work in a series of hitherto unpublished letters*, ed. Francis Darwin and A. C. Seward (London: Murray, 1903), vol. 1, p. 55.

22. Darwin, *Life and Letters*, vol. 1, p. 88; and idem, *More Letters*, vol. 1, p. 408.

23. Charles Darwin and Alfred Russel Wallace, *Charles Darwin and Alfred Russel Wallace: Evolution by Natural Selection* [a transcript of the 1842 Sketch and 1844 Essay], ed. Gavin de Beer (Cambridge: Cambridge University Press, 1958).

24. Charles Darwin, *Charles Darwin's Natural Selection: being the second part of his big species book written from 1856 to 1858*, ed. R. C. Stauffer (Cambridge: Cambridge University Press, 1975).

25. Charles Darwin, *On the Origin of Species by means of Natural Selection*, facsimile edition, with an introduction by Ernst Mayr (Cambridge, Mass.: Harvard University Press, 1964), pp. 365–80, especially pp. 366, 372.

26. Joseph Hooker, *Flora Novae Zelandiae* (London: Reeve Brothers, 1853–1855).

27. Darwin and Wallace, *Evolution by Natural Selection*, pp. 67–72.

28. Lyell, *Principles*, vol. 2, pp. 141, 169–171, 173, 176.

29. Darwin and Wallace, *Evolution by Natural Selection*, pp. 180–81.

30. Darwin and Wallace, *Evolution by Natural Selection*, pp. 181–82.

31. Darwin, *Natural Selection*, p. 557.

32. Darwin, *Natural Selection*, p. 551.

33. Janet Browne, "The Charles Darwin–Joseph Hooker correspondence: an analysis of manuscript resources and their use in biography," *Journal of the Society for the Bibliography of Natural History* 8 (1978): 353.

34. Darwin, *Natural Selection*, pp. 558–60.

35. Joseph Hooker, "Note on the replacement of species in the Colonies and elsewhere," *Natural History Review* (1864): 123–27; Lucile H. Brockway, *The Role of the British Royal Botanic Gardens* (New York: Academic Press, 1979); and Greta Jones, *Social Darwinism and English thought: the interaction between biological and social theory* (Sussex: Harvester Press, 1980).

36. Darwin, *Natural Selection*, editorial comments, pp. 528–34.

37. Archives, Royal Botanic Gardens, Kew, JDH Antarctic Letters 1839–1843, letter dated 1 June 1842, Falkland Islands, to Charles Lyell, the elder.

38. Joseph Hooker, "Outlines of the distribution of Arctic plants," *Transactions of the Linnean Society of London* 23 (1862): 251–348.

39. Hooker, *Flora Novae Zelandiae*, p. xxiv (fn.); and Darwin Archive, Cambridge University Library Archives, Dar 100, p. 109 *et seq*. Hooker provided a map, which is not reproduced in Darwin, *Natural Selection* with the rest of Hooker's comments, pp. 575–77. See also Hooker, *Flora Novae Zelandiae*, pp. xxiii–xxv.

40. Darwin, *More Letters*, vol. 1, p. 439.

41. Darwin, *More Letters*, vol. 1, p. 438.

42. Darwin Archive, Cambridge University Library Archives, Dar 100, pp. 111, 162.

43. Darwin, *More Letters*, vol. 1, p. 437.

44. Darwin, *Life and Letters*, vol. 2, p. 87.

45. Darwin, *More Letters*, vol. 1, p. 437.

CHAPTER SIX: A KEYSTONE TO CREATION

1. Darwin Archives, Cambridge University Library Archives, Dar 114, letter 27, dated 10 February 1845.

2. Louis Agassiz, "On the succession and development of organic beings," *Edinburgh New Philosophical Journal* 33 (1842): 388–389.

3. Bert Hanson, "The early history of glacial theory in British geology," *Journal of Glaciology* 9 (1970): 135–41; F. J. North, "Centenary of the glacial theory," *Proceedings of the Geologists' Association* 54 (1943): 1–28; and M. J. S. Rudwick, "The glacial theory," *History of Science* 8 (1970): 137–57.

4. Louis Agassiz, *Contributions to the Natural History of the United States of America* (Boston: Little, Brown & Co., 1857), vol. 1, "Essay on Classification," and reprinted in Guy Davenport, *The Intelligence of Louis Agassiz: A specimen book of his scientific writings* (Boston: Beacon Press, 1963), pp. 31–120.

5. Davenport, *Intelligence of Louis Agassiz*, pp. 31–39.

6. Davenport, *Intelligence of Louis Agassiz*, p. 69.

7. Davenport, *Intelligence of Louis Agassiz*, p. 68.

8. Davenport, *Intelligence of Louis Agassiz*, pp. 68–69.

9. Davenport, *Intelligence of Louis Agassiz*, p. 68.

10. Edward Forbes's letter to Leonard Horner is bound into Horner's copy of Forbes's "On the light thrown on geology by submarine researches," delivered to the Royal Institution of Great Britain, 23 February 1844, and printed in the *Edinburgh New Philosophical Journal* 36 (1844): 318–27. Horner's copy is in the British Museum (Natural History) General Library L60F(3).

11. E. Forbes, "Report on the Mollusca and Radiata of the Aegean Sea, and on their distribution, considered as bearing on geology," *Reports of the British Association for the Advancement of Science* 1843 (1844): 130–93.

12. E. Forbes, *Outlines of the natural history of Europe. The natural history of the European seas . . . edited and continued by R. Godwin-Austen* (London: Van Voorst, 1859), pp. 80–81.

13. Forbes, "Mollusca and Radiata," p. 171.

14. As in E. Forbes, "On the connexion between the distribution of the existing fauna and flora of the British Isles, and the geological changes which have affected their area, especially during the epoch of the northern drift," *Memoirs of the Geological Survey of Great Britain* 1 (1846): 336–432.

15. Forbes, "On the connexion," p. 351.

16. Johann Reinhold Forster, *Observations made during a voyage round the world, on physical geography, natural history, and ethnic philosophy* (London: n.p., 1778), p. 174.

17. Hewett Cottrell Watson, "Observations made during the summer of 1832 on the temperature and vegetation of the Scottish Highland mountains," *Edinburgh New Philosophical Journal* 14 (1833): 317–23.

18. James Wilson, "Remarks on the distribution of insects and account of Coleoptera taken in Sutherland in June 1834," *Edinburgh New Philosophical Journal* 14 (1835): 292–300.

19. Theodore Cantor, "Latitudinal distribution of reptiles inhabiting the Malayan peninsula and islands and other localities," *Edinburgh New Philosophical Journal* 44 (1848): 271–86.

20. William Thompson, "On the physical and geographical distribution of the birds of Ireland," *Edinburgh New Philosophical Journal* 46 (1849): 264–76.

21. Théodore Lacordaire, "On the geographical distribution of insects," *Edinburgh New Philosophical Journal* 28 (1839–1840): 182–83.

22. Henri Milne-Edwards, "Scientific Intelligence, Zoology," *Edinburgh New Philosophical Journal* 22 (1837): 398.

23. E. Forbes, "On the question in Natural History, have genera, like species, centres of distribution," *Athenaeum* 1062 (4 March 1848): 247, and reprinted in "Scientific Intelligence," *Edinburgh New Philosophical Journal* 45 (1848): 175–76.

24. E. Forbes, "Presidential address delivered at the Anniversary meeting of the Geological Society of London," *Quarterly Journal of the Geological Society* 10 (1854): xix–lxxxi.

25. E. Forbes, "On the manifestation of polarity in the distribution of organised beings in time," *Notices of the Proceedings of the Royal Institution of Great Britain* 1 (1854): 428–33.

26. Forbes, "Presidential address," p. lxiii; and "On the manifestation of polarity," p. 431, for a similar expression.

27. Forbes, "On the manifestation of polarity," p. 432.

28. G. L. L. Buffon, *Barr's Buffon: Buffon's Natural History containing a theory of the earth* (London: J. S. Barr, 1792), vol. 4.

29. James C. Prichard, *Researches into the physical history of Man*, ed. G. W. Stocking (Chicago: University of Chicago Press, 1973), p. 100.

30. William Swainson, *A Treatise on the Geography and Classification of Animals* (London: Lardner's Cabinet Cyclopedia, 1835).

31. Joseph Hooker, *Flora Novae Zelandiae* (London: Reeve Bros., 1853–1855), p. xxvii.

32. J. D. Hooker, "On the vegetation of the Galapagos Archipelago, as compared with that of some other tropical islands and of the continent of America," *Transactions of the Linnean Society of London* 20 (1851): 258. The short quotation preceding this one is from p. 236.

33. Hooker, "On the vegetation of the Galapagos," p. 259.

34. J. Hooker, "Lecture on Insular Floras, delivered before the British Association for the Advancement of Science at Nottingham, August 27, 1866," *Journal of Botany* 5 (1867): 23–31.

35. Alphonse de Candolle, *Géographie botanique raisonnée, ou exposition des faits principaux et des lois concernant la distribution géographique des plantes de l'époque actuelle* (Paris: Masson, 1855), pp. 1334–35.

36. Joseph Hooker, "Géographie botanique raisonnée . . . par M. Alph. de Candolle; A Review," *Hooker's Kew Journal of Botany* 8 (1856): 54–64, 82–88, 112–21, 151–57, 181–91, 214–19, 248–56, and separately printed in London, 1856, pp. 48–49.

37. Darwin Archive, Cambridge University Library Archives, Dar 100, p. 43 (not dated, but probably 1845).

38. Darwin Archive, Cambridge University Library Archives, Dar 100, p. 12, dated 5 April 1844.

39. Paul Farber, "The type concept in zoology during the first half of the 19th century," *Journal of the History of Biology* 10 (1977): 45–64; and Dov Ospovat, *The development of Darwin's theory: Natural history, natural theology and natural selection 1838–1859* (Cambridge: Cambridge University Press, 1981), pp. 115–45.

40. J. D. Hooker and T. Thomson, *Flora Indica. Volume 1 with an Introductory Essay*, 1 vol. (London: W. Pamplin, 1855), pp. 30–31 (fn.); and J. Hooker, "On the Cedars of Lebanon, Taurus, Algeria and India," *Natural History Review* (1862): 11–18.

41. Darwin Archive, Cambridge University Library Archives, Dar 100, p. 12, dated 5 April 1844.

42. Charles Darwin, *More Letters of Charles Darwin; a record of his work in a series of hitherto unpublished letters*, ed. Francis Darwin and A. C. Seward (London: Murray, 1903), vol. 1, p. 437, letter from Hooker to Darwin. Darwin's thesis of representative species was contained in a chapter on geographical distribution for his "big species book," transcribed in R. C. Stauffer, ed., *Charles Darwin's Natural Selection: being the second part of his big species book written from 1856 to 1858* (Cambridge: Cambridge University Press, 1975), pp. 528–66, 577–80.

CHAPTER SEVEN: ORGANIC CHANGE

1. Alfred Russel Wallace, *My Life: A record of events and opinions* (London: Chapman and Hall, 1905), especially chap. 6; and John R. Durant, "Scientific naturalism and social reform in the thought of A. R. Wallace," *British Journal for the History of Science* 12 (1979): 31–58.

2. William Swainson, *A Treatise on the Geography and Classification of Animals* (London: Lardner's Cyclopedia, 1835), pp. 1–18.

3. W. Swainson, *Geography and Classification*, p. 9.

4. Wallace's copy of Swainson is in the Linnean Society Library; his comment comes from p. 9. It is also described by H. Lewis McKinney, *Wallace and Natural Selection* (New Haven: Yale University Press, 1972), p. 6.

5. Robert Chambers, *Vestiges of the natural history of Creation*, facsimile edition with an introduction by Gavin de Beer (Leicester: Leicester University Press, 1969), p. 154.

6. A. R. Wallace, *The Wonderful Century: Its successes and its failures* (London: Sonnenschein, 1898), pp. 137–38.

7. Chambers, *Vestiges*, p. 219.

8. Chambers, *Vestiges*, pp. 151, 160.

9. Wallace, *My Life*, vol. 1, p. 254, letter dated 28 December 1846.

10. A. R. Wallace, *A Narrative of Travels on the Amazon and Rio Negro*

(London: Reeve & Co., 1853), pp. 472–75. For other discussions of geographical segregation in the Amazon basin, see McKinney, *Wallace and Natural Selection*, pp. 13–26; and Wilma George, *Biologist Philosopher: A study of the life and writings of A. R. Wallace* (London: Abelard-Schuman, 1964).

11. Charles Darwin, *The Life and Letters of Charles Darwin, including an Autobiographical chapter*, ed. Francis Darwin (London: Murray, 1887), vol. 2, p. 380.

12. Wallace, *My Life*, vol. 1, p. 358, letter to H. W. Bates dated 4 January 1858.

13. A. R. Wallace, "On the law which has regulated the introduction of new species," *Annals and Magazine of Natural History* 16, no. 2 (1855): 184–96, and reprinted in *A Delicate Arrangement: The Strange Case of Charles Darwin and Alfred Russel Wallace*, by A. C. Brackman (New York: Times Books, 1980), pp. 311–26. My citations are from Brackman.

14. Wallace, "On the law," p. 352.

15. Wallace, "On the law," p. 316.

16. Wallace, *My Life*, vol. 1, p. 355.

17. Charles Lyell, *Sir Charles Lyell's Scientific Journals on the Species Question*, ed. Leonard C. Wilson (New Haven: Yale University Press, 1970), p. 3.

18. Lyell, *Scientific Journals*, p. 3.

19. Lyell, *Principles of Geology, being an attempt to explain the former changes of the earth's surface by reference to causes now in operation* (London: Murray, 1830–1833), vol. 2, chaps. 1–11, inclusive; and Michael J. Bartholomew, "Lyell and evolution: an account of Lyell's response to the prospect of an evolutionary ancestry for Man," *British Journal for the History of Science* 6 (1973): 261–303.

20. Darwin, *Life and Letters*, vol. 1, p. 84; vol. 2, pp. 67–71; and *Darwin's Journal*, ed. Gavin de Beer, *Bulletin of the British Museum (Natural History)*, Historical ser. 2, pt. 1 (1959): 14. See also Lyell's account of the meeting in Lyell, *Scientific Journals*, pp. 54–55.

21. Wallace's letter no longer exists, although its contents can be inferred from Darwin's (belated) reply, reproduced in James Marchant, *Alfred Russel Wallace: Letters and reminiscences* (London: Cassell, 1916), vol. 1, pp. 129–31. See also Barbara G. Beddall, "Wallace, Darwin, and the theory of natural selection: A study in the development of ideas and attitudes," *Journal of the History of Biology* 1 (1968): 289–98.

22. Marchant, *Letters and reminiscences*, vol. 1, p. 130; also printed in Darwin, *Life and Letters*, vol. 2, pp. 95–96.

23. Marchant, *Letters and reminiscences*, vol. 1, p. 67, letter dated 4 January 1858.

24. A. R. Wallace, "Note on the theory of permanent and geographical varieties," *The Zoologist* 16 (1858): 5887–88.

25. Wallace, "Permanent and geographical varieties," p. 5888.

26. Charles Darwin, *The Voyage of the Beagle*, introd. H. G. Cannon (London: Dent, Everyman, 1959), pp. 166–67. This point has been independently noted in Beddall, "Wallace, Darwin, and the theory of natural selec-

tion," p. 297. See also McKinney, *Wallace and Natural Selection*, pp. 160–71, for Wallace's recollections of his recognition of natural selection in Malthus.

27. Charles Darwin and Alfred Russel Wallace, "On the tendency of species to form varieties, and on the perpetuation of varieties and species by natural means of selection," *Journal of the Linnean Society of London (Zoology)* 3 (1858): 45–62, reprinted in C. Darwin and A. R. Wallace, *C. R. Darwin and A. R. Wallace: Evolution by natural selection* [a transcript of Darwin's *Sketch* and *Essay*, and the Linnean Society papers, 1 July 1858], ed. Gavin de Beer (Cambridge: Cambridge University Press, 1958), pp. 255–79. My quote comes from pp. 270–71.

28. Darwin and Wallace, *Evolution by natural selection*, p. 274.

29. Charles Darwin, *More letters of Charles Darwin, a record of his work in a series of hitherto unpublished letters*, ed. Francis Darwin and A. C. Seward (London: Murray, 1903), vol. 1, p. 117, letter to Leonard Horner, 29 August 1844.

30. Charles Lyell and Roderick Impey Murchison, "On the excavation of valleys, as illustrated by the volcanic rocks of central France," *Edinburgh New Philosophical Journal* 12 (1829): 15–48; and Lyell's review of George Poulett Scrope, *Memoir on the geology of central France, including the volcanic formations of Auvergne, the Velay and the Vivarais* (London: Longman, 1827), in the *Quarterly Review* 36 (1827): 437–83. See also M. J. S. Rudwick, "Lyell on Etna, and the antiquity of the earth," in *Toward a History of Geology*, ed. Cecil J. Schneer (Cambridge, Mass.: M.I.T. Press, 1969), pp. 288–304.

31. Charles Darwin, "Geological notes made during a survey of the east and west coasts of S. America, in the years 1832, 1833, 1834 and 1835, with an account of a transverse section of the Cordilleras of the Andes between Valparaiso and Mendoza," *Proceedings of the Geological Society of London* 2 (1836): 210–12. Communicated by Adam Sedgwick and read on 18 November 1835. See also C. Darwin, *Extracts of letters from C. Darwin, Esq., to Prof. Henslow. Printed for private distribution* (Cambridge: Cambridge University Press for the Cambridge Philosophical Society, 1835). For further reprintings of these letters and others written during the Beagle voyage see Charles Darwin, *Darwin and Henslow: The growth of an idea. Letters 1831–1860*, ed. Nora Barlow (London: Murray, 1967).

32. Charles Darwin, "Observations of proofs of recent elevation on the coast of Chile, made during the survey of H.M.S. *Beagle*, commanded by Capt. Fitzroy, R.N.," *Proceedings of the Geological Society of London* 2 (1837): 449.

33. Darwin, *Life and Letters*, vol. 1, p. 70. My point is exemplified by the title of Darwin's first exposition of his coral reef theory: "On certain areas of elevation and subsidence in the Pacific and Indian oceans, as deduced from the study of coral formations," *Proceedings of the Geological Society of London* 2 (1837): 552–54.

34. Darwin, *Voyage of the Beagle*, p. 379. For a general account of the Galapagos flora and fauna see pp. 363–83.

35. Frank J. Sulloway, "Geographic Isolation in Darwin's thinking: The vicissitudes of a crucial idea," *Studies in History of Biology* 3 (1979): 26–27; and Sandra Herbert, "The place of man in the development of Darwin's theory of transmutation. Part I. To July 1837," *Journal of the History of Biology* 7 (1974): 244, 255–256.

36. Charles Darwin, "The Red notebook of Charles Darwin," ed. Sandra Herbert, *Bulletin of the British Museum (Natural History)*, Historical ser. 7 (1980): 1–164.

37. C. Darwin, "Observations of proofs of recent elevation on the coast of Chile, made during the survey of H.M.S. *Beagle*, commanded by Capt. Fitzroy, R.N.," *Proceedings of the Geological Society of London* 2 (1837): 446–49; idem, "A sketch of the deposits containing extinct Mammalia in the neighbourhood of the Plata," *Proceedings of the Geological Society of London* 2 (1837): 542–44; idem, "On certain areas of elevation and subsidence in the Pacific and Indian oceans, as deduced from the study of coral formations," *Proceedings of the Geological Society of London* 2 (1837): 552–54; idem, "On the connexion of certain volcanic phaenomena, and on the formation of mountain-chains and volcanoes as the effects of continental elevations," *Proceedings of the Geological Society of London* 2 (1838): 654–60.

38. The notebooks themselves are housed in the Darwin Archive, Cambridge University Library Archives, Dar 121–Dar 127 (notebooks A to E, M and N), except for the "RN" notebook kept at Down House, Kent. Published versions are available: Sandra Herbert, ed., "The Red notebook of Charles Darwin," *Bulletin of the British Museum (Natural History)*, Historical ser. 7 (1980): 1–164; Gavin de Beer, ed., "Darwin's notebooks on the transmutation of species," *Bulletin of the British Museum (Natural History)*, Historical ser. 2, pts. 2–6 (1960–1961), Historical ser. 3, pt. 5 (1967); Howard E. Gruber and Paul H. Barrett, *Darwin on Man: A psychological study of scientific creativity together with Darwin's early and unpublished notebooks* (London: Wildwood House, 1974). A comprehensive edition of the notebooks is in preparation at Cambridge, *Darwin's Theoretical Notebooks*, ed. P. Barrett, S. Herbert, D. Kohn, and S. Smith. "New" notebooks are described in a forthcoming paper, D. Kohn, S. Smith, and R. Stauffer, "New light on *The Foundations of the Origin of Species*: Darwin's 1839–1844 Notebooks and Essays, a reconstruction of the archival record," *Journal of the History of Biology* (1982).

39. The edition Darwin read was the sixth: T. R. Malthus, *An essay on the principles of population: or, a view of its past and present effects on human happiness with an inquiry into our prospects respecting the future removal or mitigation of the evils which it occasions*, 6th ed. (London: Murray, 1826). The most helpful accounts of Darwin's discovery of Malthus's theorem are found in David Kohn, "Theories to work by: rejected theories, reproduction and Darwin's path to natural selection," *Studies in History of Biology* 4 (1980): 67–170; Sydney Smith, "The origin of the *Origin* as discerned from Charles Darwin's notebooks and his annotations in the books he read between 1837 and 1842," *Advancement of Science* 16 (1960): 391–401; Robert M. Young, "Malthus and the evolutionists: the common context of biological and social

theory," *Past and Present* 43 (1969): 109–45; Silvan S. Schweber, "The origin of the *Origin* revisited," *Journal of the History of Biology* 10 (1977): 229–316. A more unconventional view is put forward by Dov Ospovat, "Darwin after Malthus," *Journal of the History of Biology* 12 (1979): 211–30.

40. Charles Darwin, *The foundations of The Origin of Species, a sketch written in 1842*, ed. Francis Darwin (Cambridge: Cambridge University Press, printed but not published, 1909).

41. Charles Darwin, *The foundations of The Origin of Species: Two essays written in 1842 and 1844*, ed. Francis Darwin (Cambridge: Cambridge University Press, 1909). This volume contains an exact duplication of the 1842 *Sketch* as published in the other 1909 work cited in n. 41 and a transcript of the 1844 *Essay*. Both *Sketch* and *Essay* were reissued as a new transcription by Gavin de Beer: *C. R. Darwin and A. R. Wallace: Evolution by natural selection* (Cambridge: Cambridge University Press, 1958), with the *Sketch* occupying pp. 39–88, the *Essay* pp. 89–254. My citations are taken from the de Beer volume.

42. Darwin and Wallace, *Evolution by natural selection*, pp. 68–69, 72.

43. Darwin and Wallace, *Evolution by natural selection*, pp. 70–71. The same view is more fully explained in Darwin's 1844 *Essay*, pp. 195–208.

44. Camille Limoges, *La Sélection naturelle* (Paris: Galien, 1973); Sandra Herbert, "Darwin, Malthus, and Selection," *Journal of the History of Biology* 4 (1971): 209–17; Michael Ruse, "Charles Darwin and artificial selection," *Journal of the History of Ideas* 36 (1975): 339–50.

CHAPTER EIGHT: DARWIN'S DISTRIBUTION

1. As reflected in Charles Darwin, *Charles Darwin's Natural Selection: being the second part of his big species book written from 1856 to 1858*, ed. R. C. Stauffer (Cambridge: Cambridge University Press, 1975). Several scholars have investigated the significant differences between versions of Darwin's later theories; see, for example, M. J. S. Hodge, "The structure and strategy of Darwin's 'Long Argument,'" *British Journal for the History of Science* 10 (1977): 237–46; and Dov Ospovat, *The development of Darwin's theory: Natural history, natural theology and natural selection, 1838–1859* (Cambridge: Cambridge University Press, 1981).

2. Charles Darwin and Alfred Russel Wallace, *C. R. Darwin and A. R. Wallace: Evolution by natural selection*, ed. by Gavin de Beer (Cambridge: Cambridge University Press, 1958), p. 201.

3. Charles Darwin, *The Life and Letters of Charles Darwin, including an Autobiographical chapter*, ed. Francis Darwin (London: Murray, 1887), vol. 2, p. 25. See also Darwin and Wallace, *Evolution by natural selection*, pp. 172–73.

4. Darwin Archive, Cambridge University Library Archives, Dar 114, p. 36, postmarked 12 July 1845.

5. Darwin Archive, Cambridge University Library Archives, Dar 100, fol. 20, undated reply to Darwin's letter of 12 July 1845. Darwin's response to Hooker's letter is in Charles Darwin, *More Letters of Charles Darwin, a*

record of his work in a series of hitherto unpublished letters, ed. Francis Darwin and A. C. Seward (London: Murray, 1903), vol. 1, p. 51.

6. Charles Darwin, *On the origin of species by means of natural selection, or the preservation of favoured races in the struggle for life* (London: Murray, 1859), p. 358. For Darwin's letters to Hooker on this subject see Darwin, *Life and Letters*, vol. 2, pp. 54–59; and Darwin, *More Letters*, vol. 1, pp. 416–20.

7. Darwin, *Origin of Species*, p. 360.

8. Darwin, *Origin of Species*, p. 385.

9. Edward Forbes, "On the connexion between the distribution of the existing fauna and flora of the British Isles, and the geological changes which have affected their area, especially during the epoch of the northern drift," *Memoirs of the Geological Survey of Great Britain* 1 (1846): 347–50.

10. Darwin Archive, Cambridge University Library Archives, Dar 205.2, unfoliated slip dated 21 April, no year. Continuing reorganisation of the Darwin manuscripts at Cambridge means that the contents of Dar 205 might be subject to some changes in foliation or pagination. For a general discussion of Darwin's view of land bridges and how this related to Wallace's theoretical views, see Martin Fichman, "Wallace, Zoogeography and the problem of landbridges," *Journal of the History of Biology* 10 (1977): 43–63.

11. See M. J. S. Rudwick, "Darwin and Glen Roy: a 'Great Failure' in scientific method?" *Studies in the History and Philosophy of Science* 5 (1974): 99–185, for a full account of the different interpretations of these shelflike structures.

12. Darwin, *Life and Letters*, vol. 2, p. 38.

13. Darwin, *Life and Letters*, vol. 2, p. 72.

14. Darwin, *Life and Letters*, vol. 2, p. 75.

15. Darwin and Wallace, *Evolution by natural selection*, pp. 95, 112–13, 114, 133–34.

16. Darwin, *Life and Letters*, vol. 2, p. 37, letter to Hooker dated 31 June 1849.

17. Darwin, *Natural Selection*, p. 90.

18. M. T. Ghiselin, *The Triumph of the Darwinian Method* (Los Angeles: University of California Press, 1969), pp. 103–86, and David Kohn, "Theories to work by: rejected theories, reproduction and Darwin's path to natural selection," *Studies in History of Biology* 4 (1980): 67–170, discuss the significance of sex and sexual reproduction for Darwin's theories.

19. R. C. Olby, "Charles Darwin's manuscript on pangenesis," *British Journal for the History of Science* 1 (1963): 251–61; and G. Geison, "Darwin and heredity: the evolution of his hypothesis of pangenesis," *Journal of the History of Medicine* 24 (1969): 375–411.

20. Peter Vorzimmer, "Charles Darwin and blending inheritance," *Isis* 54 (1963): 371–390; and Vorzimmer's *Charles Darwin: the years of controversy, 1859–1882* (Philadelphia: Temple University Press, 1970).

21. Charles Darwin, *Darwin's Journal*, ed. Gavin de Beer, *Bulletin of the British Museum (Natural History)*, Historical ser. 2, pt. 1 (1959): 13.

22. For a full account of Darwin's arithmetical studies see Janet Browne,

"Darwin's botanical arithmetic and the 'Principle of Divergence,' 1854–1858," *Journal of the History of Biology* 13 (1980): 53–89.

23. Darwin's copy of Alphonse de Candolle, *Géographie botanique raisonnée, ou exposition des faits principaux et des lois concernant la distribution géographique des plantes de l'époque actuelle* (Paris: Masson, 1855), housed at Cambridge University Library Archives, p. 528, with Darwin's slip pasted in.

24. Darwin Archive, Cambridge University Library Archives, Dar 15.2, fol. 4.

25. Darwin, *Natural Selection*, p. 533.

26. Darwin, *Natural Selection*, Chapter XI, "Geographical Distribution," pp. 528–66.

27. Darwin Archive, Cambridge University Library Archives, Dar 205.9, unfoliated slip dated December 1855.

28. Darwin Archive, Cambridge University Library Archives, Dar 205.5, unfoliated slip dated February 1855.

29. Darwin Archive, Cambridge University Library Archives, Dar 205.5, unfoliated slip dated November 1854. In his recent book Dov Ospovat has discussed the same 1854–1855/6 theory, although he emphasised somewhat different points from those in my account. See Ospovat, *Development of Darwin's theory*, pp. 87–114, 170–90.

30. Charles Lyell, *Sir Charles Lyell's Scientific Journals on the Species Question*, ed. Leonard C. Wilson (New Haven: Yale University Press, 1970), p. 54.

31. Ernst Mayr, *Evolution and the diversity of life: Selected essays* (Cambridge, Mass.: Harvard University Press, Belknap Press, 1976), especially in "Darwin and Isolation," pp. 120–28. Mayr's view has been furthered by F. J. Sulloway, "Geographic isolation in Darwin's thinking: The vicissitudes of a crucial idea," *Studies in History of Biology* 3 (1979): 23–65; Malcolm J. Kottler, "Charles Darwin's biological species concept and theory of geographical speciation: The transmutation notebooks," *Annals of Science* 35 (1978): 275–97; and R. Alan Richardson, "Biogeography and the genesis of Darwin's ideas on transmutation," *Journal of the History of Biology* 14 (1981): 1–41.

32. Finished on 13 October 1856, according to *Darwin's Journal*, p. 14.

33. Darwin, *Natural Selection*, p. 252.

34. Darwin, *Natural Selection*, p. 252.

35. Darwin, *Natural Selection*, p. 252.

36. Darwin, *More Letters*, vol. 1, p. 99, dated 22 August [1857]. For a full account of the limited range of contemporary material relating to Darwin's discovery of a "principle," see Browne, "Darwin's botanical arithmetic."

37. Charles Darwin and Alfred Russel Wallace, "On the tendency of species to form varieties, and on the perpetuation of varieties and species by natural means of selection," *Journal of the Linnean Society of London (Zoology)* 3 (1858): 45–62, and reprinted in Darwin and Wallace, *Evolution by natural selection*, pp. 255–79. Darwin's extract from his letter to Gray is reprinted on pp. 264–67.

38. Darwin, *Life and Letters*, vol. 2, p. 125, dated 5 September [1857], and in Darwin and Wallace, *Evolution by natural selection*, p. 266.

39. Darwin, *Origin of Species*, p. 112.

40. Darwin, *Life and Letters*, vol. 2, p. 104. This letter has, of course, been noted by historians, although no one has as yet decided what it meant to either Lubbock or Darwin. See, for example, Fred Somkin, "The contributions of Sir John Lubbock, F.R.S., to the *Origin of Species*," *Notes and Records of the Royal Society* 17 (1962): 183–91.

41. Darwin, *Life and Letters*, vol. 2, pp. 103–04.

42. See, for instance, Darwin, *Life and Letters*, vol. 2, p. 104, for an example of the "new" Lubbock technique. Darwin immediately tried one or two calculations on previously worked material, as in the Darwin Archive, Cambridge University Library Archives, Dar 16.2, fol. 243. Only five days after the conversation with Lubbock, Darwin had gone through his old calculations on C. C. Babington, *Manual of British Botany*, 3rd ed. (London: Van Voorst, 1851), correcting and marking these as "useless."

43. Darwin, *Life and Letters*, vol. 1, p. 84.

44. Comprehensively described in D. Ospovat, *Development of Darwin's theory*, pp. 115–69. Other scholars have written on the division of labour and its meaning for Darwin: Camille Limoges, "Darwin, Milne-Edwards et le principe de divergence," *Actes XIIe Congres Internationale d'Histoire des Sciences* 8 (1968): 111–15; and Silvan S. Schweber, "Darwin and the political economists: Divergence of character," *Journal of the History of Biology* 13 (1980): 195–289.

45. Darwin, *Life and Letters*, vol. 2, p. 105.

46. Darwin, *Darwin's Journal*, p. 14.

CHAPTER NINE: CONCLUSION

1. Ronald Good, *The Geography of the Flowering Plants*, 4th ed. (London: Longman, 1974); and idem, "Plant Geography," in *A Century of Progress in the Natural Sciences, 1853–1953* (San Francisco: California Academy of Sciences, 1955). See also Nicholas Polunin, *Introduction to Plant Geography and some related sciences* (London: Longman, 1960); P. J. Darlington, "Darwin and zoogeography," *Proceedings of the American Philosophical Society* 103 (1959): 307–19; and Wilma George, *Biologist Philosopher: A study of the life and writings of A. R. Wallace* (London: Abelard-Schuman, 1964). For a slightly different view see W. B. Turrill, *Joseph Dalton Hooker: Botanist, Explorer and Administrator* (London: The Scientific Book Club, 1963). Turrill suggests that modern evolutionary biogeography began with Joseph Hooker.

2. Martin Rudwick has described the same kind of phenomenon in "Darwin and Glen Roy: A 'Great Failure' in scientific method?" *Studies in the History and Philosophy of Science* 5 (1974): 99–185, where Darwin, Agassiz, and other geologists all "saw" the topography of the glens of Lochabar in different ways. John Dean, "Controversy over classification: a case study from the history of botany," in *Natural Order: Historical Studies of Scientific Cul-*

ture, ed. Barry Barnes and Steven Shapin (Beverly Hills, Calif.: Sage Publications, 1979), describes another instance of different perceptions in late-nineteenth- and twentieth-century botanical classification. Interestingly, Dean shows how conflict between taxonomists has not led to consensus: a situation somewhat similar to the case described in the present book. The way that these perceptions can be regulated by social groupings is discussed in Rudwick's "Charles Darwin in London: The integration of public and private science," *Isis* 267 (1982): 186–206, and, from another angle, by Gerald Geison, "Scientific change, emerging specialities, and research schools," *History of Science* 19 (1981): 20–40.

3. Bentley Glass, "The establishment of modern genetical theory as an example of the interaction of different models, techniques and inferences," in *Critical problems in the history of science*, ed. Marshall Clagett (Madison: University of Wisconsin Press, 1959); and Elizabeth B. Gasking, "Why was Mendel's work ignored?" *Journal of the History of Ideas* 20 (1959): 60–84.

4. G. Lemaine, R. MacLeod, M. Mulkay, and P. Weingart, eds., *Perspectives on the emergence of scientific disciplines* (The Hague: Mouton, 1976), especially the introduction, pp. 1–23.

Bibliography

Manuscripts are not listed in this bibliography, but the collections I have referred to are as follows:

Cambridge University Library, Charles Darwin Archive, containing the vast majority of Darwin's papers, his correspondence, library, and reprint collection. I refer to item or box numbers as described in the *Handlist of Darwin papers at the University Library Cambridge* (Cambridge: University Library, 1960) and the supplementary list available at the library.

Archives of the Royal Botanic Gardens, Kew, which includes Joseph Hooker's papers, except for those letters to Darwin housed at the Cambridge University Library, and other material such as that by H. C. Watson.

Imperial College Archives, Imperial College, London, which holds a fine collection of the manuscripts of T. H. Huxley and Edward Forbes.

British Museum (Natural History), manuscript collection.

PRIMARY SOURCES

Achenwall, G. *Staatsverfassung der heutigen vornehmsten Europäischen Reiche*. 5th ed. Gottingen: Vandenhoeck, 1768.

de Acosta, J. *Historia natural y moral de las Indias*. Seville: J. de Lyon, 1590.

Agassiz, L. "Essay on Classification." In *Contributions to the Natural History of the United States of America*, vol. 1. Boston: Little, Brown & Co., 1857.

———. "On the succession and development of organic beings." *Edinburgh New Philosophical Journal* 33 (1842): 388–99.

———. "The primitive diversity and number of animals in geological times." *Edinburgh New Philosophical Journal* 57 (1854): 271–92.

———. *Studies on Glaciers*. Translated by A. V. Carrozzi. New York: Hafner, 1967.

Brayley, E. W. "On the odour exhaled from certain organic remains in the Diluvium of the Arctic circle." *Philosophical Magazine* 9 (1831): 411–18.

Brerewood, E. *Enquiries touching the diversity of languages and religions through the chief parts of the world*. London: J. Bill, 1614.

Brongniart, Adolphe. "General considerations on the nature of the vegetation which covered the surface of the earth, at the different epochs of the formation of its crust." *Edinburgh New Philosophical Journal* 6 (1828–1829): 349–71.

———. "On the great divisions of the vegetable kingdom occurring in the different geological formations." *Edinburgh New Philosophical Journal* 40 (1846): 285–87.

———. *Prodrome d'une histoire des végétaux fossiles*. Paris: Levrault, 1828.

Brown, R. *General Remarks, geographical and systematical, on the botany of Terra Australis*. London: n.p., 1814.

von Buch, C. L. *Description physique des iles Canaries, suivie d'une indication des principaux volcans du globe*. Translated by C. Boulanger. Paris: n.p., 1836.

Buffon, G. L. L. *Barr's Buffon: Buffon's Natural History containing a theory of the earth*. London: J. S. Barr, 1792.

de Candolle, Alphonse. *Géographie botanique raisonnée, ou exposition des faits principaux et des lois concernant la distribution géographique des plantes de l'époque actuelle*. Paris: Masson, 1855.

———. "On the history of fossil vegetables." *Edinburgh New Philosophical Journal* 18 (1834): 81–102.

de Candolle, Augustin. "Géographie botanique." In *Dictionnaire des sciences naturelles*, edited by F. C. Levrault, vol. 18, pp. 359–436. Paris: Levrault, 1820. Printed separately as *Essai élémentaire de géographie botanique*. Paris and Strasbourg: n.p., 1820.

———. "Mémoire sur la géographie des plantes de France, considerée dans ses rapports avec la hauteur absolue." *Mémoires de physique et de chimie de la Société d'Arcueil* 3 (1817): 262–322.

———. *Projet d'un flore physico-géographique de la vallée du Leman*. Geneva: Paschoud, 1821.

———. *Theorie élementaire de la botanique*. 2nd ed. Paris: Roret, 1819.

Cantor, T. "Latitudinal distribution of reptiles inhabiting the Malayan peninsula and islands and other localities." *Edinburgh New Philosophical Journal* 44 (1848): 271–86.

Chambers, R. *Vestiges of the natural history of Creation*. Facsimile edition, with an introduction by G. de Beer. Leicester: Leicester University Press, 1969.

Clift, W., and R. Jameson. "Report by Mr Clift in regard to the fossil bones found in the Bone Caves and Bone-Breccia of New Holland." *Edinburgh New Philosophical Journal* 10 (1831): 393–96.

——— and W. Parish. "An account of the discovery of portions of three skeletons of the Megatherium in the province of Buenos Ayres." *Proceedings of the Geological Society of London* 1 (1832): 403–04.

Cuvier, G. "Mémoire sur les espèces d'elephans vivantes et fossiles." *Mémoires de l'Institut National des Sciences et Arts (Classe Mathématiques et Physiques)* 2 (1799): 1–22.

———. *Recherches sur les ossemens fossiles de quadrupèdes, où l'on rétablit les caractères de plusieurs espèces d'animaux que les révolutions du globe paroissent avoir détruites.* Paris: Chez Deterville, 1812.

——— and Alexandre Brongniart. *Essai sur la géographie minéralogique des environs de Paris.* Paris: Baudouin, 1811.

Dana, J. "On the geographical distribution of Zoophytes." *Edinburgh New Philosophical Journal* 43 (1847): 41–45.

Darwin, C. *Charles Darwin's Natural Selection: being the second part of his big species book written from 1856 to 1858.* Edited by R. C. Stauffer. Cambridge: Cambridge University Press, 1975.

———. *Darwin and Henslow: The growth of an idea. Letters 1831–1860.* Edited by N. Barlow. London: Murray, 1967.

———. *Darwin's Journal.* Edited by G. de Beer. *Bulletin of the British Museum (Natural History).* Historical ser. 2, pt. 1 (1959): 1–21.

———. "Darwin's notebooks on the transmutation of species." Edited by G. de Beer. *Bulletin of the British Museum (Natural History).* Historical ser. 2, pts. 2–6 (1960–1961), and ser. 3, pt. 5 (1967).

———. *Extracts of letters from C. Darwin, Esq., to Professor Henslow. Printed for private distribution.* Cambridge: Cambridge University Press for the Cambridge Philosophical Society, 1835.

———. *The foundations of The Origin of Species: Two essays written in 1842 and 1844.* Edited by F. Darwin. Cambridge: Cambridge University Press, 1909.

———. "Geological notes made during a survey of the east and west coasts of S. America, in the years 1832, 1833, 1834 and 1835, with an account of a transverse section of the Cordilleras of the Andes between Valparaiso and Mendoza." *Proceedings of the Geological Society of London* 2 (1836): 210–12.

———. *The Life and Letters of Charles Darwin, including an Autobiographical chapter.* Edited by F. Darwin. London: Murray, 1887.

———. *More Letters of Charles Darwin, a record of his work in a series of hitherto unpublished letters.* Edited by F. Darwin and A. C. Seward. London: Murray, 1903.

———. "Observations of proofs of recent elevation on the coast of Chile, made during the survey of H.M.S. *Beagle*, commanded by Capt. Fitzroy, R.N.," *Proceedings of the Geological Society of London* 2 (1837): 446–49.

———. "On certain areas of elevation and subsidence in the Pacific and Indian oceans, as deduced from the study of coral formations." *Proceedings of the Geological Society of London* 2 (1837): 552–54.

———. "On the connexion of certain volcanic phænomena, and on the formation of mountain-chains and volcanoes as the effects of continental elevations." *Proceedings of the Geological Society of London* 2 (1838): 654–60.

———. *On the Origin of Species by means of Natural Selection.* Facsimile edition, with an introduction by E. Mayr. Cambridge, Mass.: Harvard University Press, 1964.

———. "The Red notebook of Charles Darwin." Edited by S. Herbert. *Bulletin of the British Museum (Natural History).* Historical ser. 7 (1980): 1–164.

———. "A sketch of the deposits containing extinct Mammalia in the neighbourhood of the Plata." *Proceedings of the Geological Society of London* 2 (1837): 542–44.

———. *The Voyage of the Beagle.* With an introduction by H. G. Cannon. London: Dent, Everyman, 1959.

——— and A. R. Wallace. *Charles Darwin and Alfred Russel Wallace: Evolution by natural selection.* Edited by G. de Beer. Cambridge: Cambridge University Press, 1958. Containing reprints of Darwin's 1842 *Sketch* and 1844 *Essay*, and Darwin and Wallace's 1858 Linnean Society paper.

———. "On the tendency of species to form varieties, and on the perpetuation of varieties and species by natural means of selection." *Journal of the Linnean Society of London (Zoology)* 3 (1858): 45–62.

Ehrenberg, C. G. *Die Infusionsthierchen als volkommene Organismen.* Leipsig: Voss, 1838.

Fleming, J. "On the value of the evidence from the animal kingdom, tending to prove that the Arctic regions formerly enjoyed a milder climate than at present." *Edinburgh New Philosophical Journal* 6 (1829): 277–86.

Forbes, E. "The life of an individual and the duration of a species." *Edinburgh New Philosophical Journal* 53 (1852): 130–35.

———. "On the connexion between the distribution of the existing fauna and flora of the British Isles, and the geological changes which have affected their area, especially during the epoch of the northern drift." *Memoirs of the Geological Survey of Great Britain* 1 (1846): 336–432.

———. "On the light thrown on geology by submarine researches." *Edinburgh New Philosophical Journal* 36 (1844): 318–27.

———. "On the manifestation of polarity in the distribution of organised beings in time." *Notices of the Proceedings of the Royal Institution of Great Britain* 1 (1854): 428–33.

———. "On the question in Natural History, have genera, like species, centres of distribution?" *Athenaeum* 1062 (March 4, 1848): 247. Reprinted in "Scientific Intelligence." *Edinburgh New Philosophical Journal* 45 (1848): 175–76.

———. *Outlines of the natural history of Europe. The natural history of the European seas . . . edited and continued by R. Godwin-Austen*. London: Van Voorst, 1859.

———. "Presidential address delivered at the Anniversary meeting of the Geological Society of London." *Quarterly Journal of the Geological Society* 10 (1854): xix–lxxxi.

———. "Report on the Mollusca and Radiata of the Aegean Sea, and on their distribution, considered as bearing on geology." *Reports of the British Association for the Advancement of Science* 1843 (1844): 130–93.

Forster, J. R. *Observations made during a voyage round the world, on physical geography, natural history, and ethnic philosophy*. London: n.p., 1778.

Fourier, J. B. "Mémoire sur le refroidissement seculaire de globe terrestre." *Bulletin de la Société Philomathique de Paris* (1820): 58–70.

Grotius, H. *De origine gentium Americanarum*. Paris and Amsterdam: n.p., 1642. Translated by E. M. Goldsmid as *On the origin of the native races of America*. In *Bibliotheca Curiosa*, edited by E. M. Goldsmid. Edinburgh: privately printed, 1883–1888.

Hale, M. *The Primitive Origination of Mankind considered and examined according to the light of Nature*. London: W. Shrowsbery, 1677.

Herschel, J. *Preliminary Discourse on the study of Natural Philosophy*. London: Lardner's Cabinet Cyclopedia, 1830.

Hooke, R. *Lectures and Discourses of Earthquakes and Subterranean Eruptions*. In *The Posthumous Works of Robert Hooke*. London: R. Waller, 1705.

Hooker, J. D. *Flora Novae Zelandiae*. London: Reeve Brothers, 1853–

1855. Part of Hooker's *Botany of the Antarctic Voyage of H.M.S. Erebus and Terror . . . 1839–1843*. London: Reeve Brothers, 1844–1860.
———. "Géographie botanique raisonnée . . . par M. Alph. de Candolle; A Review." *Hooker's Kew Journal of Botany* 8 (1856): 54–64, 82–88, 112–21, 151–57, 181–91, 214–19, 248–56. Separately printed in London, 1856.
———. "Lecture on Insular Floras, delivered before the British Association for the Advancement of Science at Nottingham, August 27, 1866." *Journal of Botany* 5 (1867): 23–31.
———. "Note on the replacement of species in the Colonies and elsewhere." *Natural History Review* (1864): 123–27.
———. "On the Cedars of Lebanon, Taurus, Algeria and India." *Natural History Review* ii (1862): 11–18.
———. "On the vegetation of the Galapagos Archipelago, as compared with that of some other tropical islands and of the continent of America." *Transactions of the Linnean Society of London* 20 (1851): 235–62.
———. "Outlines of the distribution of Arctic plants." *Transactions of the Linnean Society of London* 23 (1862): 251–348.
——— and T. Thomson. *Flora Indica. Volume I with an Introductory Essay*. 1 vol. London: W. Pamplin, 1855.
Horner, L. "Anniversary Address of the President." *Quarterly Journal of the Geological Society* 2 (1846): 145–221.
———. "Anniversary Address of the President." *Quarterly Journal of the Geological Society* 3 (1847): xxii–xc.
Hornius, G. *Arca Noae sive historia imperiorum et regnorum a condito orbe ad nostra tempora*. Leyden: n.p., 1666.
———. *De originibus Americanis libri quatuor*. Leyden: n.p., 1652.
von Humboldt, A. *Aspects of Nature in different lands and different climates with scientific elucidations*. Translated by Mrs Sabine. London: Longman, 1849.
———. *Cosmos: A Sketch of a physical description of the Universe*. Translated by E. C. Otté. London: Bohn, 1849–1858.
———. *Essai sur la géographie des plantes; accompagne d'un tableau physique des régions equinoxiales*. Paris: Levrault, 1805. Facsimile edition of the first part only by the Society for the Bibliography of Natural History, Sherborn Fund Facsimiles 1 (1959).
———. *Florae Fribergensis specimen*. Berlin: Rottman, 1793.
———. "Géographie des plantes." In "Géographie botanique," by Augustin de Candolle, vol. 18, pp. 422–32. In *Dictionnaire des sciences naturelles*, edited by F. C. Levrault. Paris: Levrault, 1820.

———. *Personal Narrative of travels to the Equinoctal regions of the New Continent, 1799–1804*. Translated by H. M. Williams. London: Longman, 1821–1825.

———. *Prolegomena*. In *Nova Genera et Species Plantarum quas in peregrinatione ad plagam eaquinoctiatem Orbis Novi collegerunt, descripserunt et adumbraterunt . . .* , by A. Bonpland, A. von Humboldt, and C. S. Kunth. Paris: Lutetiae, 1815–1825.

Hutton, J. *Theory of the Earth, with proofs and illustrations*. Edinburgh: Cadell and Davies, 1795.

Illiger, K. "Tabellarische Ubersicht uber die Vertheilung der Vogel uber die Erde." *Abhandlungen der Königlichen Akademie der Wissenschaften zu Berlin* (1812–1813): 221–36.

Jameson, R. "On the fossil flora." *Edinburgh New Philosophical Journal* 13 (1832): 349–50.

Kircher, A. *Arca Noë in tres libros digesta*. Amsterdam: Waesberg, 1675.

Lacordaire, T. "On the geographical distribution of insects." *Edinburgh New Philosophical Journal* 27 (1838–1839): 94–111, 333–51 and 28 (1839–1840): 170–83, 224–38.

Laetius, J. *Notae ad dissertationem H. Grotii de origine gentium Americanarum*. Paris and Amsterdam: n.p., 1643.

de La Peyrère, I. *Praeadamitae* . . . London: n.p., 1655. Translated as *Men before Adam, or a discourse upon the 12th, 13th and 14th verses of the Fifth Chapter of the Epistle of the Apostle Paul to the Romans*. London: n.p., 1656.

Latreille, P. A. "Introduction a la géographie générale des Arachnides et des Insectes." *Edinburgh New Philosophical Journal* 5 (1821): 370–78 and 6 (1822): 51–62.

Le Clerc, J. *Commentarii philologici et paraphrases in vet Testam*. Amsterdam: n.p., 1690–1731.

Lindley, J. "On the mode of determining fossil plants." *Edinburgh New Philosophical Journal* 13 (1832): 221–28.

Linnaeus, C. *Museum S.R.M. Adolphi Friderici Regis*. Holmiae: P. Momma, 1754. Translated by J. E. Smith as *Reflections on the study of Nature*. London: George Nicol, 1786.

———. *Oratio de Telluris habitabilis incremento* . . . Leyden: C. Haak, 1744. Translated by F. J. Brand in *Select dissertations from the Amonitates Academicae: A Supplement to Mr Stillingfleet's Tracts relating to Natural History*. London: G. Robinson, 1781.

———. *Systema Naturae 1735, facsimile edition with a translation of the Observationes*. Edited by M. S. J. Engel-Ledeboer and H. Engel. Nieuwkoop: n.p., 1964.

Lipsius, J. *Physiologiae stoicorum. Libri III* in his *Opera*. Antwerp: Moreti, 1637.

Lyell, C. *Principles of Geology, being an attempt to explain the former changes of the earth's surface by reference to causes now in operation*. London: Murray, 1830–1833.

———. "Review of George Poulett Scrope, *Memoir on the geology of central France, including the volcanic formations of Auvergne, the Velay and the Vivarais* (London: Longman, 1827)." *Quarterly Review* 36 (1827): 437–83.

———. *Sir Charles Lyell's Scientific Journals on the Species Question*. Edited by L. C. Wilson. New Haven: Yale University Press, 1970.

——— and R. I. Murchison. "On the excavation of valleys, as illustrated by the volcanic rocks of central France." *Edinburgh New Philosophical Journal* 12 (1829): 15–48.

Malthus, T. R. *An essay on the principles of population: or, a view of its past and present effects on human happiness with an inquiry into our prospects respecting the future removal or mitigation of the evils which it occasions*. 6th ed. London: Murray, 1826.

Meyen, F. J. F. *Outlines of the geography of plants with particular enquiries concerning . . . the principal cultivated plants on which the prosperity of nations is based*. Translated by M. Johnson. London: Ray Society, 1846.

Milne-Edwards, H. "Scientific Intelligence, Zoology." *Edinburgh New Philosophical Journal* 22 (1837): 397–99.

Murchison, R. I. *The geology of Russia in Europe and the Ural mountains*. London: Murray, 1845.

———. "Presidential address." *Reports of the British Association for the Advancement of Science* 1846 (1847): xxvii–xlii.

Murray, A. "Thoughts regarding the influence of rock on native vegetables." *Edinburgh New Philosophical Journal* 11 (1831): 56–66.

van der Myl [Mijl], A. *De origine animalium et migratione populorum*. Geneva: n.p., 1667.

d'Orbigny, A. "General considerations regarding the palaeontology of South America compared with that of Europe." *Edinburgh New Philosophical Journal* 35 (1843): 362–72.

Owen, R. *On the Nature of Limbs*. London: J. van Voorst, 1849.

Pentland, J. "On the fossil bones of Wellington valley, New Holland or New South Wales." *Edinburgh New Philosophical Journal* 12 (1832): 301–08.

Piso, G., and G. Marcgravi. *Historia Naturalis Brasiliae*. Amsterdam: Elzevier, 1648.

Poole, M. *Synopsis Criticorum aliorumque Sacrae Scripturae interpretum.* London: n.p., 1669–1676.

Prichard, J. C. "On the relations of Ethnology to other branches of knowledge." *Edinburgh New Philosophical Journal* 43 (1847): 307–36.

———. *Researches into the physical history of Man.* Edited by G. W. Stocking. Chicago: University of Chicago Press, 1973.

Ray, J. *Miscellaneous Discourses concerning the dissolution and changes of the world.* London: S. Smith, 1692.

von Schlotheim, E. F. "Abhandlung uber die Krauter-Abdrücke im Schieferthon und sandstein der Steinkohlen-Formationen." *Hoffmagazin* 1 (1801): 76–95.

———. *Beschreibung merkwurdiger Krauter-Abdrücke und Pflanzen-Versteinerungen.* Gotha: Becker, 1804.

Sedgwick, A. "Anniversary Address." *Proceedings of the Geological Society of London* 1 (1830): 187–212.

———. "Anniversary Address." *Proceedings of the Geological Society of London* 1 (1831): 281–97.

Sprengel, K. P. J., and Augustin de Candolle. *Elements of the Philosophy of Plants containing the Principles of Scientific Botany.* Edinburgh: W. Blackwood, 1821.

Stillingfleet, E. *Origines Sacrae, or a rational account of the grounds of Christian faith.* London: H. Mortlock, 1662.

Swainson, W. *A Treatise on the Geography and Classification of Animals.* London: Lardner's Cabinet Cyclopedia, 1835.

Thompson, W. "On the physical and geographical distribution of the birds of Ireland." *Edinburgh New Philosophical Journal* 46 (1849): 264–76.

"A uniformity of climate prevailed over the earth prior to the time of the Deluge." *Edinburgh New Philosophical Journal* 8 (1830): 366–67.

Wagner, R. "The numerical relations of animals from Linnaeus to the present day." *Edinburgh New Philosophical Journal* 15 (1833): 221–27.

Wallace, A. R. *My Life: A record of events and opinions.* London: Chapman and Hall, 1905.

———. *A Narrative of Travels on the Amazon and Rio Negro.* London: Reeve & Co., 1853.

———. "Note on the theory of permanent and geographical varieties." *The Zoologist* 16 (1858): 5887–88.

———. "On the law which has regulated the introduction of new species." *Annals and Magazine of Natural History* 16, no. 2 (1855):

184–96. Reprinted in *A Delicate Arrangement: The Strange Case of Charles Darwin and Alfred Russel Wallace*, by A. C. Brackman. New York: Times Books, 1980.

———. *The Wonderful Century: Its successes and its failures.* London: Sonnenschein, 1898.

Watson, H. C. *Cybele Britannica: or British plants and their geographical relations.* London: Longman, 1847–1859; Supplement, 1860; Compendium, 1868–1870.

———. "Observations made during the summer of 1832 on the temperature and vegetation of the Scottish Highland mountains in connection with their height above the sea." *Edinburgh New Philosophical Journal* 14 (1833): 317–23.

Whewell, W. *History of the Inductive Sciences, from the earliest to the present times.* London: Parker, 1837.

———. *The Philosophy of the Inductive Sciences, founded upon their history.* 2nd ed. London: Parker, 1847.

Whiston, W. *A New Theory of the Earth.* London: n.p., 1696.

White, G. *The Natural History and Antiquities of Selborne: The standard edition by E. T. Bennett.* Revised edition by J. E. Harting. London: Swan Sonnenschein, 1889.

Willdenow, K. *Herbarium Willdenow: Alphabetical Index.* Edited by P. Hiepko. Switzerland: Interdocumentation Company, 1972.

———. *Principles of Botany.* Edinburgh: Blackwood, 1805.

Wilson, J. "Remarks on the distribution of insects and account of Coleoptera taken in Sutherland in June 1834." *Edinburgh New Philosophical Journal* 14 (1835): 292–300.

Winch, N. *An Essay on the geographical distribution of plants.* Newcastle: Walker, 1819.

Woodward, J. *An Essay toward a natural history of the Earth . . . with an account of the universal Deluge.* London: n.p., 1695.

Zimmermann, E. A. W. *Specimen Zoologiae Geographicae Quadrupedum domicilia et migrationes sitens.* Leyden: T. Haak, 1777.

SECONDARY SOURCES

Allen, D. C. *The Legend of Noah: Renaissance Rationalism in Art, Science and Letters.* Urbana: Illinois University Press, 1949.

Allen, D. E. *The Naturalist in Britain: A Social History.* Harmondsworth: Penguin Books, 1978.

Altick, R. D. *Victorian People and Ideas.* New York: Norton, 1973.

Andrews, H. N. *The Fossil Hunters: In Search of Ancient Plants.* Ithaca: Cornell University Press, 1980.

Bartholomew, M. J. "Lyell and evolution: an account of Lyell's response to the prospect of an evolutionary ancestry for Man." *British Journal for the History of Science* 6 (1973): 261–303.

Beddall, B. G. "Wallace, Darwin, and the theory of natural selection: A study in the development of ideas and attitudes." *Journal of the History of Biology* 1 (1968): 261–318.

Bodenheimer, F. S. "Zimmermann's *Specimen Geographicae Quadrupedum*, a remarkable zoogeographical publication of the end of the 18th century." *Archives Internationale d'Histoire des Sciences* 8 (1955): 351–57.

Bowler, P. J. *Fossils and Progress: Paleontology and the idea of a progressive evolution in the nineteenth century.* New York: Science History Publications, 1976.

Brockway, L. H. *The Role of the British Royal Botanic Gardens.* New York: Academic Press, 1979.

Browne, J. "The Charles Darwin–Joseph Hooker correspondence: an analysis of manuscript resources and their use in biography." *Journal of the Society for the Bibliography of Natural History* 8 (1978): 351–66.

———. "Darwin's botanical arithmetic and the 'Principle of Divergence,' 1854–1858." *Journal of the History of Biology* 13 (1980): 53–89.

Cannon, S. F. *Science in Culture: The early Victorian period.* London and New York: Dawson, 1978.

Coleman, W. *Georges Cuvier, Zoologist: A Study in the History of Evolutionary Theory.* Cambridge, Mass.: Harvard University Press, 1964.

Conry, Y. *Correspondence entre Charles Darwin et Gaston de Saporta.* Paris: Galien, 1972.

Crosland, M. *The Society of Arcueil: A view of French science at the time of Napoleon I.* London: Heinemann, 1967.

Cullen, M. J. *The Statistical Movement in early Victorian Britain.* Sussex: Harvester Press, 1975.

Darlington, P. J. "Darwin and zoogeography." *Proceedings of the American Philosophical Society* 103 (1959): 307–19.

Davenport, G. *The Intelligence of Louis Agassiz: A specimen book of his scientific writings.* Boston: Beacon Press, 1963.

Dean, J. "Controversy over classification: a case study from the history of botany." In *Natural Order: Historical Studies of Scientific Culture,* edited by B. Barnes and S. Shapin. Beverly Hills, Calif.: Sage Publications, 1979.

Desmond, A. "Designing the Dinosaur: Richard Owen's response to Robert Edmond Grant." *Isis* 70 (1979): 224–34.

Durant, J. R. "Scientific naturalism and social reform in the thought of A. R. Wallace." *British Journal for the History of Science* 12 (1979): 31–58.

Egerton, F. N. "Ancient Sources for Animal Demography." *Isis* 59 (1968): 175–89.

———. "A Bibliographical guide to the history of general Ecology and Population Ecology." *History of Science* 15 (1977): 189–215.

Farber, P. "The type concept in zoology during the first half of the nineteenth century." *Journal of the History of Biology* 10 (1977): 45–64.

Fichman, M. "Wallace, Zoogeography and the problem of land-bridges." *Journal of the History of Biology* 10 (1977): 43–63.

Fieldler, H., S. Scheibe, and E. Germer. *Georg Forster, Naturforscher, Weltreisender, Humanist und Revolutionar*. Wörlitz: Staatliche Schlosser, 1975.

Fitter, R. S. R. *The Ark in our Midst: The story of the introduced animals of Britain*. London: Collins, 1959.

Foucault, M. *The Order of Things: An Archaeology of the Human Sciences*. London: Tavistock, 1970.

Gasking, E. B. "Why was Mendel's work ignored?" *Journal of the History of Ideas* 20 (1959): 60–84.

Geison, G. "Darwin and heredity: the evolution of his hypothesis of pangenesis." *Journal of the History of Medicine* 24 (1969): 375–411.

———. "Scientific change, emerging specialities, and research schools." *History of Science* 19 (1981): 20–40.

George, W. *Biologist Philosopher: A study of the life and writings of Alfred Russel Wallace*. London: Abelard-Schuman, 1964.

———. "Sources and background to discoveries of new animals in the sixteenth and seventeenth centuries." *History of Science* 18 (1980): 79–104.

Gerstner, P. A. "The reaction to James Hutton's use of heat as a geological agent." *British Journal for the History of Science* 5 (1971): 353–62.

Ghiselin, M. T. *The Triumph of the Darwinian Method*. Los Angeles: University of California Press, 1969.

Glass, B. "The establishment of modern genetical theory as an example of the interaction of different models, techniques and inferences." In *Critical problems in the history of science*, edited by M. Clagett. Madison: University of Wisconsin Press, 1959.

Glass, D. V. *Numbering the People: The 18th century population controversy and the development of census and vital statistics in Britain*. London: D. C. Heath, 1973.

———. "Some aspects of the development of demography." *Journal of the Royal Society of Arts* 104 (1956): 854–69.

Good, R. *The Geography of the Flowering Plants.* 4th ed. London: Longman, 1974.

———. "Plant Geography." In *A Century of Progress in the Natural Sciences, 1853–1953*, edited by E. L. Kessel. San Francisco: California Academy of Sciences, 1955.

Gruber, H. E., and P. H. Barrett. *Darwin on Man: A psychological study of scientific creativity together with Darwin's early and unpublished notebooks.* London: Wildwood House, 1974.

Hanson, B. "The early history of glacial theory in British geology." *Journal of Glaciology* 9 (1970): 135–41.

Herbert, S. "Darwin, Malthus, and Selection." *Journal of the History of Biology* 4 (1971): 209–17.

———. "The place of man in the development of Darwin's theory of transmutation. Part I. To July 1837." *Journal of the History of Biology* 7 (1974): 217–58.

Hill, C. *The Century of Revolution, 1603–1714.* London: Nelson, 1980.

———. *The Intellectual Origins of the English Revolution.* Oxford: Oxford University Press, 1965.

———. *Society and Puritanism in Pre-Revolutionary England.* London: Secker and Warburg, 1964.

Hodge, M. J. S. "The structure and strategy of Darwin's 'Long Argument.'" *British Journal for the History of Science* 10 (1977): 237–46.

Hooykaas, R. *Religion and the rise of modern science.* Edinburgh: Scottish Academic Press, 1972.

Huxley, T. H. "Memoir of Professor Edward Forbes, F.R.S." *Literary Gazette* 1975 (25 November 1854): 1016–18.

Jaccard, P. "Sur le coéfficient generique." *Chronica Botanica* 6 (1941): 361–64, 389–91.

Jones, G. *Social Darwinism and English thought: the interaction between biological and social theory.* Sussex: Harvester Press, 1980.

Jordanova, L. J., and R. Porter, eds. *Images of the Earth: Essays in the History of the Environmental Sciences.* Chalfont St. Giles, Bucks.: British Society for the History of Science Monographs, 1979.

Kohn, D. "Theories to work by: rejected theories, reproduction and Darwin's path to natural selection." *Studies in History of Biology* 4 (1980): 67–170.

———, S. Smith, and R. Stauffer. "New light on *The Foundations of the Origin of Species*: Darwin's 1839–1844 Notebooks and Essays, a reconstruction of the archival record." *Journal of the History of Biology* 1982, in press.

Kottler, M. J. "Charles Darwin's biological species concept and theory of geographical speciation: The transmutation notebooks." *Annals of Science* 35 (1978): 275–97.

Lemaine, G., R. MacLeod, M. Mulkay, and P. Weingart, eds. *Perspectives on the emergence of scientific disciplines.* The Hague: Mouton, 1976.

Lenoir, T. "The Göttingen School and the development of transcendental Naturphilosophie in the Romantic era." *Studies in History of Biology* 5 (1981): 111–205.

Lepenies, W. "Linnaeus's *Nemesis divina* and the concept of divine retaliation." *Isis* 266 (1982): 11–27.

Limoges, C. "Darwin, Milne-Edwards et le principe de divergence." *Actes XIIe Congres Internationale d'Histoire des Sciences* 8 (1968): 111–15.

———. *La Sélection naturelle.* Paris: Galien, 1973.

McKinney, H. L. *Wallace and Natural Selection.* New Haven: Yale University Press, 1972.

Marchant, J. *Alfred Russel Wallace: Letters and reminiscences.* London: Cassell, 1916.

Mayr, E. *Evolution and the diversity of life: Selected essays.* Cambridge, Mass.: Harvard University Press, Belknap Press, 1976.

Merz, J. T. *A History of European Scientific Thought in the Nineteenth Century.* New York: Dover, 1965.

Meyer-Abich, A. *Alexander von Humboldt.* Bonn: Internationes [The Alexander von Humboldt Foundation], 1969.

Morton, A. G. *History of Botanical Science.* London: Academic Press, 1981.

Nelson, G. "From Candolle to Croizat: Comments on the history of biogeography." *Journal of the History of Biology* 11 (1978): 269–305.

North, F. J. "Centenary of the glacial theory." *Proceedings of the Geologists' Association* 54 (1943): 1–28.

Olby, R. C. "Charles Darwin's manuscript on pangenesis." *British Journal for the History of Science* 1 (1963): 251–61.

Ospovat, D. "Darwin after Malthus." *Journal of the History of Biology* 12 (1979): 211–30.

———. *The development of Darwin's theory: Natural history, natural theology and natural selection, 1838–1859.* Cambridge: Cambridge University Press, 1981.

———. "Lyell's theory of Climate." *Journal of the History of Biology* 10 (1978): 317–39.

Pearson, E. S., and M. G. Kendall, eds. *Studies in the History of Statistics and Probability*. London: Charles Griffin, 1970.

Pearson, K. *The History of Statistics in the seventeenth and eighteenth centuries against the changing background of intellectual, scientific and religious thought: Lectures given by Karl Pearson at University College during the academic sessions 1921–1933*. Edited by E. S. Pearson. London: Griffin, 1978.

Polunin, N. *Introduction to Plant Geography and some related sciences*. London: Longman, 1960.

Porter, R. S. "Charles Lyell and the principles of the history of geology." *British Journal for the History of Science* 9 (1976): 91–103.

———. *The Making of Geology: Earth Science in Britain, 1660–1815*. Cambridge: Cambridge University Press, 1972.

Raven, C. E. *English Naturalists from Neckham to Ray: A Study of the Making of the Modern World*. Cambridge: Cambridge University Press, 1947.

———. *Natural Religion and Christian Theology*. Cambridge: Cambridge University Press, 1953.

Reilly, P. C. *Athanasius Kircher, S.J.: Master of a Hundred Arts, 1602–1680*. Wiesbaden: Edizioni del Mondo, 1974.

Richardson, R. A. "Biogeography and the genesis of Darwin's ideas on transmutation." *Journal of the History of Biology* 14 (1981): 1–41.

Rudwick, M. J. S. "Charles Darwin in London: the integration of public and private science." *Isis* 267 (1982): 186–206.

———. "Charles Lyell's dream of a statistical palaeontology." *Palaeontology* 21 (1978): 225–44.

———. "Darwin and Glen Roy: a 'Great Failure' in scientific method?" *Studies in the History and Philosophy of Science* 5 (1974): 99–185.

———. "The glacial theory." *History of Science* 8 (1970): 137–57.

———. "Historical analogies in the geological work of Charles Lyell." *Janus* 64 (1977): 89–108.

———. "Lyell on Etna, and the antiquity of the earth." In *Toward a History of Geology*, edited by C. J. Schneer. Cambridge, Mass.: M.I.T. Press, 1969.

———. *The Meaning of Fossils: Episodes in the History of Palaeontology*. London: MacDonald, 1972.

———. "The strategy of Lyell's *Principles of Geology*." *Isis* 61 (1970): 5–33.

———. "Transposed concepts from the human sciences in the early work of Charles Lyell." In *Images of the Earth: Essays in the History of the Environmental Sciences*, edited by L. J. Jordanova and R. Porter. Chalfont St. Giles, Bucks.: British Society for the History of Science Monographs, 1979.

———. "A visual language for geology." *History of Science* 14 (1976): 149–95.

Ruse, M. "Charles Darwin and artificial selection." *Journal of the History of Ideas* 36 (1975): 339–50.

Russell, C. A., ed. *Science and Religious Belief: A selection of recent historical studies.* London: University of London Press, 1973.

Russell, E. S. *Form and Function: A Contribution to the History of Animal Morphology.* London: Murray, 1916.

Sarjeant, W. A. S., and J. B. Delair. "An Irish naturalist in Cuvier's laboratory: The letters of Joseph Pentland, 1820–1832." *Bulletin of the British Museum (Natural History).* Historical ser. 6, no. 7 (1980): 245–319.

Schweber, S. S. "Darwin and the political economists: Divergence of character." *Journal of the History of Biology* 13 (1980): 195–289.

———. "The origin of the *Origin* revisited." *Journal of the History of Biology* 10 (1977): 229–316.

Smith, S. "The origin of the *Origin* as discerned from Charles Darwin's notebooks and his annotations in the books he read between 1837 and 1842." *Advancement of Science* 16 (1960): 391–401.

Somkin, F. "The contributions of Sir John Lubbock, F.R.S., to the *Origin of Species.*" *Notes and Records of the Royal Society* 17 (1962): 183–91.

Stearn, W. T. "Humboldt's *Essai sur la Geographie des Plantes.*" *Journal of the Society for the Bibliography of Natural History* 3, part 7 (1960): 351–57.

Sulloway, F. J. "Geographic Isolation in Darwin's thinking: The vicissitudes of a crucial idea." *Studies in History of Biology* 3 (1979): 23–65.

Sweet, P. R. *Wilhelm von Humboldt: A Biography.* Columbus: Ohio State University Press, 1978.

Théodoridès, J. "Humboldt and England (1769–1859)." *British Journal for the History of Science* 3 (1966): 39–55.

Turrill, W. B. *Joseph Dalton Hooker: Botanist, Explorer and Administrator.* London: The Scientific Book Club, 1963.

Vorzimmer, P. "Charles Darwin and blending inheritance." *Isis* 54 (1963): 371–90.

———. *Charles Darwin: the years of controversy, 1859–1882.* Philadelphia: Temple University Press, 1970.

Westergaard, H. *Contributions to the History of Statistics.* The Hague and Paris: Mouton, 1969.

Westfall, R. S. *Science and Religion in Seventeenth-Century England.* New Haven: Yale University Press, 1958.

Wulff, E. V. *An introduction to historical plant geography* . . . Translated by E. Brissenden. Waltham, Mass.: Chronica Botanica Co., 1943.

Young, R. M. "Malthus and the evolutionists: the common context of biological and social theory." *Past and Present* 43 (1969): 109–45.

Index